The Illustrated Toyota Production System

A Lean Transformation Primer

By Ritsushi Tsukuda

Translated by Mark T. Nagai

Originally published as *zukai de wakaru Seisan no jitsumu – toyota seisan houshiki*

Copyright © 2006 by Ritsushi Tsukuda

Original Japanese edition published by JMA Management Center Inc.

English translation rights arranged with JMA Management Center Inc.

English translation copyright © 2008 Gemba Press

Translation by Mark T. Nagai

Edited by Gemba Press

First Edition, First Print on August 1, 2006

ISBN: 0-9786387-6-X

All rights reserved. No part of this book may be reprinted or reproduced or utilized in any form or by any electronic, mechanical, or other means, now known or hereafter invented, including photocopying and recording, or in any information storage or retrieval system, without permission in writing from the publishers.

Book and cover layout designed by Michael Wang

Cover art by Michael Wang

Printed and bound in the United States of America. Printed on acid-free paper.

Direct all inquiries to:

Gemba Press

13000 Beverly Park Road, Suite B

Mukilteo, WA 98725 USA

Phone: 1.888.Go.Gemba

Fax: 1.425.348.7234

Email: press@gemba.com

Web: www.gemba.com/press

Table of Contents

Publisher's Foreword — i
Author's Preface — ii
Translator's Preface — iv

Section 1:

The Toyota Production System (TPS) Structure and Philosophy

1. High Quality and High Profit by Eliminating Waste — 2
2. The Secret of Toyota's High Profit is their Philosophy of Making Things — 4
3. Thorough Elimination of "*3 mu*" to Build a High Quality, High Profit System — 6
4. True Efficiency vs. Apparent Efficiency — 8
5. The Basic Philosophy about Waste at Toyota — 10
6. Toyota's Seven Types of Waste — 12
7. The Steps for Eliminating of Seven Types of Waste — 14
8. How to Remove the Waste of Overproduction — 16
9. Is Inventory Evil? — 18
10. Why Do Companies Not Succeed at Zero Inventory Operation? — 20
11. Zero Inventory Exposes Internal Problems — 22
12. Zero Inventory Begins by Improving Inventory Accuracy — 24
13. To Improve Inventory Taking Accuracy — 26
14. Case Example of Inventory Taking — 28
15. The Overall Picture of TPS — 30

Section 2:

Just In Time

16. What is Just In Time?	34
17. *Heijunka* as a Prerequisite of Just In Time	36
18. Why Do Companies Not Succeed at Just In Time?	38
19. Making Processes Flow	40
20. What is Production Lead Time?	42
21. The Meaning of One-Piece Flow Production	44
22. What is the Required Takt Time for Production?	46
23. Withdrawal (Pull) by the Downstream Processes	48
24. What is the *Kanban* System?	50
25. The Types of *Kanban*	52
26. Enabling the Use of *Kanban*	54
27. Why Do We Practice 5S?	56
28. Sort is to Throw Away Unnecessary Things	58
29. Straighten is to Make Things Immediately Available	60
30. Sweep is to Focus and Thoroughly Implement	62
31. Key Points to Start Implementing TPS with 5S	64

Section 3:

Jidoka, Or Automation with Human Intelligence (Autonomation)

32. What is *Jidoka*?	68
33. The Relationship between *Jidoka* and Just In Time	70
34. Why Do Companies Not Succeed at *Jidoka*?	72
35. Quality is Built in at the Process	74
36. Differences between Work Standards and Standard Work	76
37. Key Points for Standard Work	78
38. The Standard Work Combination Sheet and the Standard Work Sheet	80
39. Reasons Why Standard Work is Difficult to Establish	82
40. The Difference between Manpower Saving and Multi-process Handling	84
41. Flexible Manpower	86
42. The Difference between Rate of Operation and Operational Availability	88
43. Key Points for Setup Time Reduction	90
44. The Steps of Setup Time Reduction	92
45. The Importance of Maintenance	94
46. Safety Takes Precedence Over Everything Else	96
Bibliography	98

Publisher's Foreword

The Toyota Production System (TPS) is now regarded worldwide as a management system and a business philosophy that combines high performance with sustainability. For most of us much remains to be learned about how to benefit from the past half-century of experimentation and advancement of Toyota's management system. *The Illustrated Toyota Production System: A Lean transformation Primer* by Ritsushi Tsukuda combines a series of 46 lessons and more than 50 charts, diagrams and tables to illustrate the Toyota Production System.

This book approaches TPS from the perspective of the overall structure and thought process of TPS, built around the two pillars of just in time and *jidoka*. The three sections of the book then link other tools and aspects of TPS to how they support a high quality, high profit production system that relentlessly drives out waste by harnessing the creativity of people. The author explains the concepts concisely, often with enough depth and insight such that I am confident the reader will have many "a-ha!" moments as they turn the pages of this book.

This book is not a step-by-step "how to" guide for TPS implementation but rather a series of reflections and insights to be aware of in advance of TPS implementation. The more practical aspects of the Toyota Production System such as *kaizen*, industrial engineering (IE) techniques, and process analysis for manpower savings are illustrated in Book2 of this series by the same author, to be published by Gemba Press in 2008.

This book is *a lean transformation primer* because it gives you a broad understanding of the principles, structures and tools that make up the Toyota Production System as well as some of the underlying thinking. The information in this book is based on the author's first-hand experiences implementing lean in Japan and will be a primer – an accelerating agent – to your lean transformation efforts.

Publisher: Jon Miller
June 2008

Author's Preface

Toyota has been receiving attention for years as a company achieving high quality and high profit, and recently seems to be at the zenith of its prosperity.

Since the burst of the economic bubble, more and more companies have shown interest in the Toyota Production System (hereafter called TPS). Those companies that have implemented it publicize their success stories, raising interest further.

Yet there seem to be many companies that cannot achieve favorable results even after implementing TPS. It is from such companies that I have started receiving requests for help.

I first encountered TPS about 33 years ago, but started to actually work with it when I attended a seminar of the Japan Management Association (JMA) instructed by a consultant, who was one of my seniors. He explained that at Toyota the concept of "overproduction is evil" emerged around 1960, and the effort since then have been to not have work in process inventories. One of the representative methods for work in process inventory reduction is the use of *kanban*. I was impressed by Toyota's system for using *kanban* to achieve business results.

Most shocking for me at the time was the management philosophy of Toyota. Having work in process inventories means that people are avoiding difficult situations, and operating without work in process inventories requires a high level of technology and management attention. I was astonished that Toyota and its group of companies had established such a mechanism over their many years.

For over 30 years I have been trying to implement TPS through various *kaizen* activities. But I have been made to realize that without strong and lasting commitment from top management, TPS will not take root and will only result in painful experiences.

On the other hand, I also encountered many successful case examples during that period. Therefore the elimination of waste has many useful aspects for any company, unless applied mistakenly. That is why this book places emphasis on waste elimination with no more than a simple explanation of the *kanban* system.

The focus of volume one is on TPS, which may seem complicated to those familiar with traditional production systems, partly because of Toyota's own jargon and terminologies, but also because of the concepts that are opposite or contradictory to existing concepts. One example is the concept that "the downstream processes withdraw materials from the upstream processes, and upstream processes make only what is withdrawn."

Such concepts are difficult to those who are used to existing production systems. Therefore, as much as possible a concise explanation is attempted for each of these concepts.

Companies like Toyota which took many years to establish their business structures encompassing people, machines, and system naturally have different steps to take for *kaizen* from those companies that have not established such structures.

The saying "not seeing the forest for the trees" describes how getting caught up in details can prevent seeing the big picture. Because Toyota is excellent at seeing the "forest," people tend to focus on that "forest" aspect. Toyota, however, is good at seeing the "trees" (details) as well and thoroughly trains their people to see the details. If companies that try to implement the TPS see the forest without seeing the trees, *kaizen* will only be superficial and short-lived.

While explaining TPS, I have included my experiences and thoughts throughout the book. It will give me great pleasure if this book can serve as a reference to those companies considering the implementation of TPS or hesitating because they feel their business structure has not yet developed to that level.

Ritsushi Tsukuda
June 2006

Translator's Preface

Is TPS applicable to only Japanese companies?

Some people ask such a question. However, there is no denying today that many companies in the world have already benefited from applying the Toyota Production System (TPS) or lean concepts to their operations. Also, it is not well known that TPS was originated from Toyota's endeavor to learn from the US, in particular the Training Within Industry (TWI) program of the US Department of War and the supermarket system and from Herry Ford's production system.

From there Toyota developed many unique and important concepts such as the 7 types of waste, takt time, flow production, downstream pull system, *kanban, heijunka* (production smoothing), *jidoka*, 5S and others. The good news is that this book explains all of these at a practical level useful for implementation.

That said it is also true that in Japan there are certain cultural foundations compatible with these practices, or rather these concepts contain certain elements of Japanese psychology. This book is ideal in demystifying such concepts. It has been a privilege to translate them from Japanese into English with the aid of my own experience with TPS, briefly described below.

How is it possible to make a car every 57 seconds?

When I visit Toyota plants I am constantly impressed by the consistency of their production speed. They maintain a smooth production flow painlessly. To the amazement of most people, one car is made at Toyota usually every 57 seconds (every 2 minutes for Lexus). In fact, at one Toyota plant you could time the speed of the production, and exactly at the speed of 57 seconds, the conveyors move the cars to be assembled from one work center to the next.

Ensuring smooth flow: Even with the occasional stoppage by the operators (who pull the string that lights up in yellow the specific work center on the *andon* board) the team leader usually fixes the tiny issues in 10 seconds or so. This becomes possible, because they work with constant steady flow based on takt time, a fundamental concept of lean manufacturing. The takt time is calculated by dividing the time available for production by the quantity needed by customer orders per day.

Stopping the flow: It is very important that operators are authorized and instructed to stop the line whenever they find any abnormality. This is a principle called autonomation. To avoid any defect, which is deviation from standard quality, is more important than achieving more efficiency.

Pleasant melodies rather than the buzzer

Insightful episodes of Mr. Taiichi Ohno, the father of TPS abound to explain the real meaning of TPS. One such is that he did not like the buzzer that was the signal given as an alert. Instead, he would recommend pleasant melodies because the buzzer would give the operators a feeling of being chased. He also suggested that such music can be chosen by the operators in the area.

Mr. Ohno taught that TPS should not be something painful for operators, for the sake of higher efficiency. He emphasized the importance of making the work enjoyable and comfortable. He explained that good *kaizen* (continuous improvement) must have harmony of:
- Production efficiency improvement,
- Safety improvement, and
- Comfort for people on the production floor.

Terminologies and translation

After some deliberation during the process of translating, it was decided to italicize all non-English words, and to:

(1) Keep certain Japanese words as they are, e.g. *kanban*
(2) Not use Japanese at all for some words
(3) Use both English and Japanese, e.g. words for 5S, such as waste (*muda*).

The first category: Certain words like *kanban* were decided to be written as they are, because the words are either already accepted or too cumbersome to replace with an English translation. It is somewhat comparable to the use of words, such as *judo, kendo*, *sushi*, or *tsunami*. For one thing, it would take extra sentences to explain what these really mean if we were to explain with English words only. It would be like calling *judo* or *kendo* simply as wrestling or fencing, just for the sake of simplification. The readers might then be misled or confused. Or, fewer people today would label *sushi* as raw fish itself which is misleading, nor tsunami as tidal wave, as discouraged by oceanographers.

The second category: If Japanese words are printed as written in the original, the readers will have to stop all the time. To avoid confusion, we dropped certain words like *muda* in many instances, although we explain somewhere else in the book that the word *muda* means waste. We don't want to prevent a smooth reading flow.

Practical Answers Illustrated for Frequently Asked Questions about TPS

As the title suggests, this book illustrates how TPS is implemented and practiced. The illustrations can be very valuable to those who try to practice TPS or lean concepts. Namely, this book provides many practical explanations of the major fundamentals of TPS and its two pillars, just in time and *jidoka* (autonomation or automation with human intelligence). These two pillars are typically illustrated in the TPS House diagram with the underlying foundation being *heijunka*, explained in section 2.

Mr. Tsukuda, lean consultant and the author of this book, answers many frequently asked questions such as "Why do companies not succeed in zero inventory operation?" "Why do companies not succeed in just in time or *jidoka*?" or "What are the reasons why standard work is hard to establish?" These fundamental and very important questions are answered with clear-cut explanations, practical illustrations and examples. Hence this book is called *The Illustrated Toyota Production System: a lean transformation primer* The readers will see many in-depth and practical answers to frequently asked questions on TPS.

As a consultant myself who has read many books on the subject both in English and Japanese and as someone who has been a lean practitioner for 20 plus years, the translation of this book has been like walking into a goldmine. In this book there are many excellent points of clarification, which you would discover only occasionally even in the real world of gemba (actual working place), let alone in books on the subject until you trained your eyes to see things through lean ways. You can develop such eyes only when you are given clear and methodical explanations based on real life examples. This book contains many pieces of practical advice with illustrations, both for beginners and advanced practitioners. I am convinced that it will enable you to gain an overall picture, plus an in-depth perspective of lean practices.

Mark Nagai, CPIM
TPS Consultant

March 2008

Section 1:

The Toyota Production System Structure and Philosophy

Lesson 1:
High Quality and High Profit by Eliminating Waste

Companies cannot continue existing without profit. Then, let us rethink what profit is. The cost composition is shown on Figure 1. Further, the composition of manufacturing cost is shown in a simplified way to make it easier to understand.

If we sell at the same price as the manufacturing cost, we cannot invest in equipment nor can we cope with the need to raise wages. It is necessary to add profit to the Total Cost (Cost of Goods Sold + Operating Expenses). This will become the selling price.

In other words, the ways to make profit can be in the following three ways:
(1) Raise the selling price
(2) Make and sell large quantities (result of mass production)
(3) Reduce cost (lower the manufacturing cost)

However, When the competition for the product is intense it is very difficult to raise the selling price. Instead we often see selling prices reduced because of competition. You can easily increase the volume of production, but with a saturated market as is often the case in today's economy, products cannot be sold. Then, the only remaining way is to lower the cost. That is why companies today are very much eager to do cost reduction.

The Toyota Production System (TPS) is characterized by its unique systems and methods for reducing not only manufacturing cost, but also through the elimination of the 7 types of waste.

Also, the TPS ensures a thorough practice of "building quality into the process." That is how they make high quality products while achieving the cost reduction.

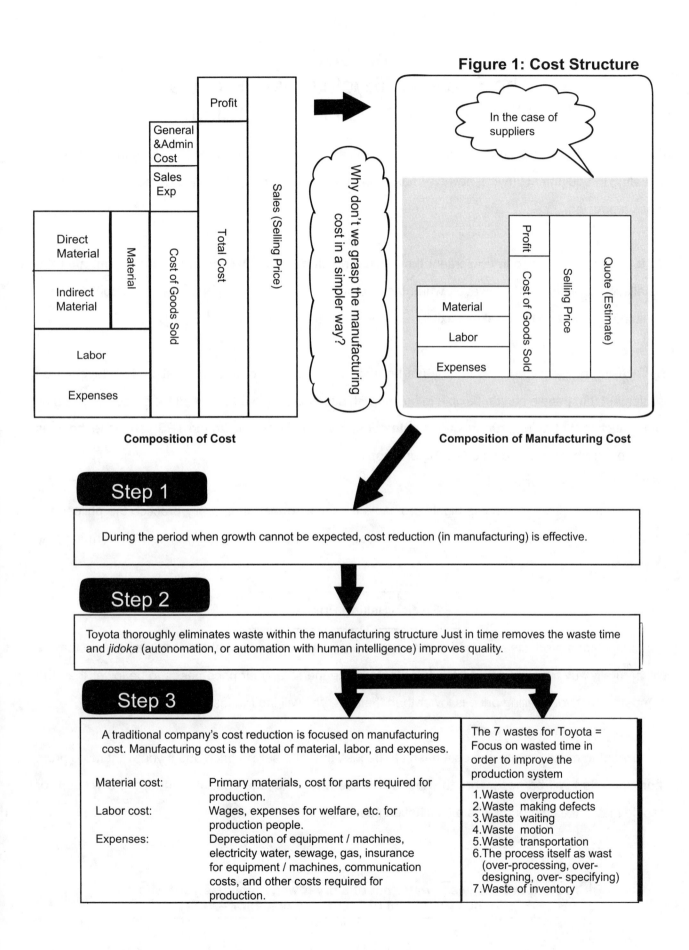

Figure 1: Cost Structure

Lesson 2:
The Secret of Toyota's High Profit is their Philosophy of Making Things

During periods when sales were brisk, companies made things by mass production focusing on investing in equipment. Toyota however has promoted "making things focusing on people, even when mass producing."

By focusing on investing in equipment the desired result can be achieved in the short term, after the installation of equipment. However, when the equipment becomes old, people tend to depend on the equipment supplier which makes flexible responses more difficult.

By focusing on people the result may not be immediate compared to the operation that depends on equipment, but people can be flexible in coping with situations such as the need for mixed-model small lot production. In the long run this will create added value increasingly. In the TPS various techniques were born from improvements centering on people.

The reason why Toyota remains highly profitable is secanse they thoroughly practice the philosophy that the profit is the selling price minus the cost, which means the selling price is determined by customers.

Profit = Selling Price – Cost

If a competitor's products are good quality and cheaper than your company's products, the customers will naturally buy from the competitor. In order to make the sale, your price needs to be lower than the competitor's. If your selling price is lower than the cost, you will incur a loss.

If you want to make a profit, the cost needs to be less than the selling price and if you want more profit thorough effort will be necessary to eliminate waste and lower the cost. It is indeed the wisdom of people that makes thorough waste elimination possible.

**You can even squeeze water out of a dry towel, if you use your wits."
Eiji Toyoda (President of Toyota Motors, 1967 – 1972)**

Figure 2: Companies with Weak Cost Awareness

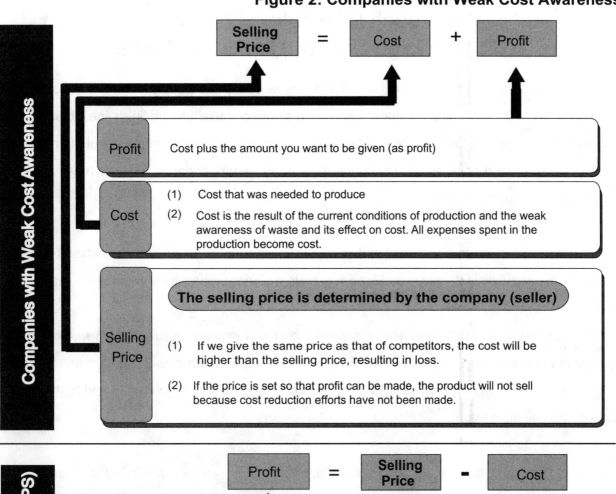

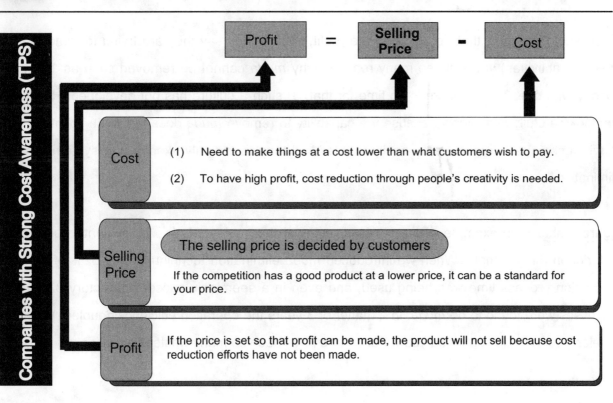

Lesson 3:
Thorough Elimination of "3 *mu*"
to Build a High Quality, High Profit System

When we watch a marathon we notice that runners who sway their bodies right and left, or hop around, seem to be OK at the beginning but many such runners cannot make it to the finish line either because of *muda* (wasteful activity) in their form or because of *muri* (physical strain) in terms of stamina. Also, those runners who go back and forth do not show good results because of *mura* (unevenness or inconsistencies in performance).

The opposite of "3 *mu*" is to be efficient and effective. Those runners who run at a consistent pace with efficient form and running style ultimately seem to have a higher probability of winning the race.

In manufacturing as well, it is possible to lower the cost and improve quality. The "3 *mu*" represents *muri* (physical strain), *muda* (wasteful activity) and *mura* (inconsistencies in performance). Toyota divided *muda* into 7 types of wastes and became an enterprise with high quality and high profit by thoroughly eliminating waste.

If you see companies that do not generate profit, even if they say they are trying to remove *muda* (wasteful activities) you still see many reasons why *muda* cannot be removed such as 1) there is nobody who does that, 2) there is no time for that, 3) such a culture has not developed. However, if you look carefully it is simply because the capability to remove *muda* does not exist. Toyota is good at this technique to make waste easily visible. And they thoroughly execute the system of waste elimination.

Figure 3 shows an example where *muda* is removed thoroughly. *Muda* was evident in all the four factories in this example. When we introduced *muda* elimination techniques only 52% of available production process time was being used, and even in a seemingly successful factory 32% of the processes were found to be *muda*. In a matter of 3 years the operation efficiency doubled. When you develop your eyes to see *muda* you will see there have been so many wasteful activities.

Figure 3: Examples of Waste Elimination

① As we searched for *muda* (wasteful activities), the operation efficiency was 52%-68% We can see that there is a lot of waste.

② If you have the techniques to find *muda*, the operation efficiency in all of the four factories will steadily rise.

③ If you do *kaizen*, the operation efficiency starts to exceed 100%.

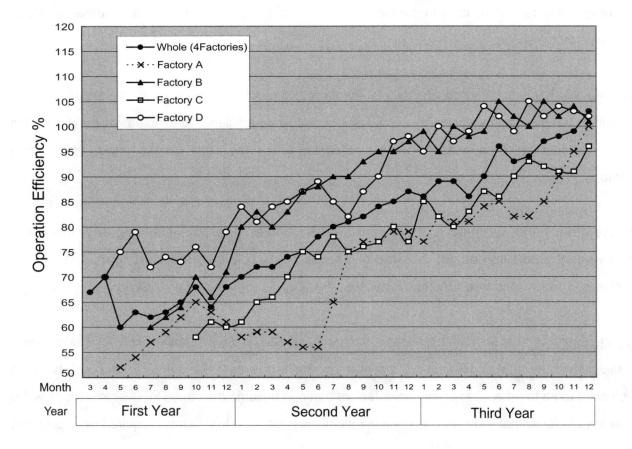

$$\text{Operation Efficiency} = \frac{\text{Process Output}}{\text{(Available production process time minus excluded production process time*)}}$$

* The time in which operation cannot be executed.

Man-Hours = (Manpower or number of persons) × (Time)

Lesson 4:
True Efficiency vs. Apparent Efficiency

Efficiency is the measurement of the effectiveness production activity. Efficiency can be expressed in various ways, but the efficiency manual work can be expressed as "**production quantity divided by number of people.**" In general, when we seek to increase efficiency we add machines or increase the number of people in order to increase the production quantity (numerator of the equation).

For example, when making 100 pieces of a certain product with 10 people, there are two ways to increase efficiency by 20%. The first is to increase the number of machines or people, which is relatively easy. The other is to decrease the number of people by making improvements, which requires creative efforts.

However, when the quantity required by a customer is 100 pieces, if you increase efficiency and make more than 100 pieces, it is waste. By improving the calculated efficiency and increasing the production quantity irrespective of the sales, the result is **apparent efficiency,** and this is something forbidden at Toyota.

On the other hand, by making 100 pieces with 8 people, whereby achieving 20% efficiency improvement production output is linked to what is being sold with the existing manpower and machines. This is called **true efficiency**, because it will lead to practical cost reduction.

During periods of rapid economic growth or when the company's sales steadily increase, waste is hard to see behind the growth. However, the present condition for most companies is that they have no such extra room to relax. One of the major factors that can become the cause for overproduction is the apparent efficiency that increases production quantity disregarding the quantity required by customers.

The improvement of **apparent efficiency** will lead to the waste of overproduction, which only increases inventory. Toyota attaches more importance to this than any other types of waste in the production gemba (actual place of work) and it is strictly forbidden.

Figure 4: True Efficiency vs. Apparent Efficiency

Efficiency = Production Quantity divided by Number of people

Current Condition:

Making 100 pieces of product with 10 people

The customer requirement is 100 pieces

Classification	Explanation
Apparent Efficiency 120 pieces made with 10 people	* 20 pieces became "waste of over-production" because the customer's required quantity was 100 pieces, which results in the "waste of inventory". * Because of 20% production increase, at the time of production, it appears to have lowered the cost. However, overproduced inventory requires extra transportation and carrying cost, which extra production outweighs the cost reduction with extra production.
True Efficiency 100 pcs made with 8 people	* Placement of production people in response to the required quantity becomes possible. * True efficiency improvement can be achieved, because it will not create unnecessary inventory, which results in cost reduction.

Toyota Language

Individual Efficiency

- By increasing the efficiency of only your own process, it appears that the efficiency of the whole production process increased as the production output increases.
- Increasing the local efficiency of each process is not desirable because it can lead to over-production and increase work in process inventory. This is harmful for production flow.

Lesson 5:
The Basic Philosophy about Waste at Toyota

Even if you are aware of the importance of waste elimination you cannot see waste unless you are trained to have eyes to see waste or have sign waste awareness. If you cannot see waste, it means that your way of looking at things is superficial. If you acquire the techniques to see things minutely, you will gradually begin to see the wastes.

You cannot look minutely at the whole, in order to do *kaizen* effectively. It is therefore important to have techniques to narrow down you focus. As a way of thinking about waste elimination methods, we separate the operations into 1) wasteful work, 2) true work, and 3) incidental work.

Wasteful work

This is work that does not add value but adds cost. This is the first type which needs to be eliminated. This includes waiting, moving things, or looking for tools. Even if you perspire a lot while moving the load, your work will be wasteful and not considered as value added

In assembly procees within a manufacturing operation you would need to do inspection after assembly, and have time for start up and ending the operation, and cleaning up after the operation, in addition to the net time assembling the component parts.

True work

This is work that is value-added, such as assembling component parts.

Incidental work

This is work that is non value-added which are necessary in addition to the true work, such as changeover. We need to exert our efforts to make changeovers as close as possible to zero.
At Toyota we say it is the managers' role to "look at their subordinates' work and identify waste and ways to make their work easier and more efficient, and to transform their motions into true work (e.g. motion itself is not work)."

Kaizen is the practice of constant improvement through many small changes, based on the creativity and ideas of people.

Figure 5: Ideas about Waste

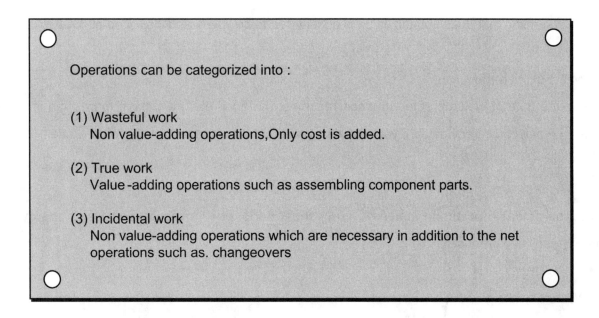
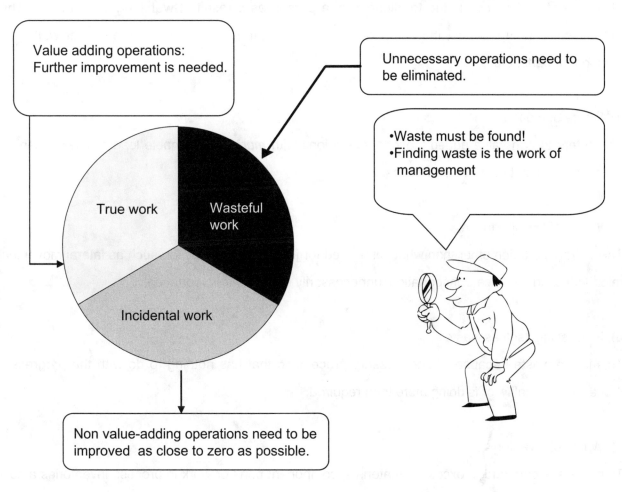

Lesson 6:
Toyota's Seven Types of Waste

In TPS, anything that does not add value is considered to be waste and classified as follows:

1) Waste of overproduction

This is the type of waste created in labor, machine, material, etc. by making more than is needed, thinking that it can be sold later. If you make things earlier than needed, it will create waste.

2) Waste of defects

This type of waste occurs in materials, component parts, rework requirements and energy lost in making defective products.

3) Waste of waiting

This type of waste occurs due to idleness of a person as a result of watching the machine while the machines is in process, the stoppage of operation during machine breakdowns, or waiting for component parts, etc.

4) Waste of motion

This refers to non value added motions, operations with *muri* (unreasonable load or overburden), or waste in inefficient postures or motions.

5) Waste of transportation

This is transportation other than what is required for just in time production such as lateral movement, reloading, long-distance transportation, unnecessarily repeated transportation.

6) Processing itself as waste

This is the waste because of unnecessary processing that has nothing to do with the progress of processes or quality, or is doing more than required.

7) Waste of inventory

This waste occurs due to excess in materials, component parts or work in process inventories among different work centers. Loss occurs due to the carrying cost of inventory.

Figure 6: The Seven Types of Waste (1 of 2)

Waste	Contents	Causes	Countermeasures
Waste Overproduction	① Excess production ② Producing too early ③ Obstacle to production flow ④ Increase in Finished Goods / Work in process inventory ⑤ Decrease of asset turnover	• Insufficient preliminary discussion with customer • Production planning based on individual's experience and intuition • Excess manpower • Excess equipment • Large lot production • Production load fluctuation • Process trouble (defects, machine breakdown)	•Thorough preliminary discussion with customer •Standardization of production planning •*Heijunka* production •One piece flow •Small lot production •Production using *kanban* •SMED or Single Minute Exchange of Dies (Changeover time reduction) •Introduction of takt time
Waste of Making Defects	① Waste of material ② Decrease in operating ratio ③ *Muda* in inspection ④ Lowering confidence level due to claims ⑤ Increased inventory ⑥ *Muda* in rework	• Lax attitude towards estimating defect occurrence • Quality not built in by the process • Inspection focused control, insufficient inspection standard (especially sensual inspection) • Insufficient training, customer's demand for excessively high quality, standard work implementation insufficient	• Quality to be built in by the processes • *Jidoka* (autono-mation), Thorough practice of management by fact (*genchi, genbutsu, genjitsu* principle) • Awareness countermeasures • Prevention by asking 5 Whys • *Pokayoke* (mistake-proofing) to implement a system to assure quality • Integrate *kaizen* and quality systems (ISO 9001)
Waste of Waiting	① Insufficient Standard Work in repetitive operations ② Supervision ③ Busywork ④ Idlenes ⑤ Extra time to spare	• Unsmooth process flow • Troubles in previous or following processes • Waiting process • Busywork • Inadequate layout of machines Unbalance of operation capacity in the process • Large lot production	• Implement *heijunka* and takt time • Efforts for smooth flow, visualizing *muda* • U-shape layout • SMED or Single Minute Exchange of Dies • Splitting operations • Prohibition of filler work while waiting • Line balance analysis • Identify busywork • Devices to detect abnormality and alarm

13

Lesson 7:
The Steps for Eliminating the Seven Types of Waste

We can begin eliminating any of the seven types of waste, but in terms of an organization's capabilities to do so, there are vast differences between a company like Toyota that has accumulated *kaizen* experiences for over 50 years and those companies that have not done so. Here, let us think of the basic sequence for companies that have lots of problems

Step 1: It is necessary to have the philosophy that vigorous control is necessary so that the **waste of overproduction** is eliminated by producing only at the required timing, just in time.

Step 2: If there are too many occurrences of defects it is difficult to avoid overproduction. **Waste of making defects** need to be eliminated by attacking the root causes of the defects.

Step 3: Making improvements to waiting is easy if the main points are understood. You can make effective use of manpower by removing the **waste of waiting**.

Step 4: The waste of motion greatly affects the reduction of man-hours. However, it is necessary to exercise caution because even after removing many small wastes identified through doing motion analysis, other problems challenges may remain.

Step 5: As for the **waste of transportation**, distance, number of times of transportation, etc. need to be improved.

Step 6: With regard to the process itself as waste, we can consider any process that will not create added value to be waste.

Step 7: By implementing the steps 1 – 6 your inventory will inevitably decrease. In other words, if people try to aim at zero inventory without addressing these problems, a lot of troubles will occur. Depending on the company's capability it may be better to do the elimination of waste of inventory as one of the last steps.

> **Kanban are signals (cards, containers or electronic) that instruct production or movement of materials.**

Figure 7: The Seven Types of Waste (2 of 2)

Waste	Contents	Causes	Countermeasures
Waste of Motion	① Non value adding motions ② Motions not based on principles of motion economy	• Unclear distinction between the machine and manpower operations • Non value added motions occur unconsciously • Inadequate layout • Inadequate trainings	• Make processes flow • U-shape machine layout • Thorough training on motion economy principles • Identify and thoroughly remove busywork • Use of Standard Work Combination Sheet • Identify Value-Addedness for tasks
Waste of Transportation	① Moving materials to the side reloading ② Empty transport ③ Damage to products during transport ④ Wasteful space ⑤ Transport distance and frequency ⑥ Increased equipment for transport	• Lack of understanding that searching or moving is not a work. • Inadequate layout • Transportation Liveliness Index is low. • Production sequence and process combination, not thoroughly studied	• The concept that transport is to be avoided • Optimum transport frequency • U-shape machine layout • Making lot sizes small • Use various conveyance systems such as water spiders • Unit transport of components • Improve Transportation Liveliness Index*
Processing Itself as Muda	① Increased manpower and man-hours, due to unnecessary processes and tasks (e.g. deburring, etc.) ② Lower productivity ③ More defects ④ Old ways being followed without improvement	• Inadequate design of processes • Insufficient analysis of operation contents • Insufficient counter measures against short-time breakdown • Insufficient PM (Preventive Maintenance) • Inadequate jigs/tools • Unsatisfactory standardization • Undeveloped operator skills • Inappropriate materials	• Review operations that had been following the past custom • Solution of problems based on go see principles • Examine inspection methods • Appropriate design of processes • Preparation of jigs • Automation with robotics • Thorough standardization • Study of materials • Properly documenting records of machine breakdowns • Train skilled PM personnel
Waste of Inventory	① Finished goods work-in-process inventory ② Inventory cost (depreciation of warehouse, transport equipment maintenance cost small transport cost tax insurance; payable interest for investment supply / consumption cost obsolescence cost, etc.) ③ Inventory hides many problems	• *Heijunka* production is not implemented • It is given that large inventory is needed for shipping preparation and delivery date control. • Inadequate machine layout • Large lot production • Production that precedes orders • Extra production when waiting	• *Heijunka* production • Awareness of what inventory is truly required (make only what is sold). • Smooth flow processes • Thorough use of *kanban* • Items and information need to be moved together. • Resolve all problems within production processes (unresolved issues need to be practically nonexistent).

*Translator's note

Transportation Liveliness Index is an industrial engineering term coined by author *Kenji Endo* and widely in use among TPS practitioners in Japan. The Transportation Liveliness Index is the degree to which products left loosely on the floor can be moved at a moment's notice. When transporting products loosely placed on the floor individually, there are four types of actions that require time and labor: 1) gathering together (putting the loosely scattered items from the floor in order, typically putting into boxes), 2) picking up the boxes, 3) lifting (e.g. raise the pallet), and 4) transporting to the destination. A product left on the floor loosely and individually (not bundled or in box), for instance, has the index 0 (not ready at all to be transported), and a product on the conveyor belt or forklift has the index 4 (fully ready to be transported).

Lesson 8:
How to Remove the Waste of Overproduction

During the mass production era, many companies did not worry about overproduction. At **Toyota overproduction is strictly prohibited and seen as the root of every kind of evil. This includes producing too early** as well.

Overproduction uses extra manpower, materials and money and leads to the cost of inventory and the possibility of making unnecessary products. If we are aiming for the Toyota Production System, doing overproduction is just like destroying the function of just in time.

In manufacturing, we often see situations where someone prepares production plans for the week based on firm orders and projected forecasts. Usually the actual status is compared to the inventory ledger, but if necessary, people go to the gemba (actual place of work) to see.

This person typically calculates the yield in his or her brain based on past experiences. When the production plans need to be changed, especially with frequent changes in yield percentage, or if the inventory accuracy is not there, everything depends on this person's capability. Such a person is often called "a genius" at work. However, if you study carefully, the focus tends to be on avoiding shortages and the plans are linked to overproduction.

If you explain that people do not have to make more than the planned quantity for the day, people do not behave that way. If the individual workload is balanced, it is fine. However, operators who already finished the quantity for the day very often make use of the extra time available in producing more by assembling parts or preparing extra material kits: this is **apparent work** or busywork.

Even if your company wants to avoid overproduction it is necessary to eliminate such factors shown on the opposite page so that *heijunka* (production smoothing) by volume and by product type mix becomes possible.

> **Heijunka is the smoothing of production by averaging both the load and mix of work over a period of time.**

Figure 8: Overproduction Occurs in These Circumstances

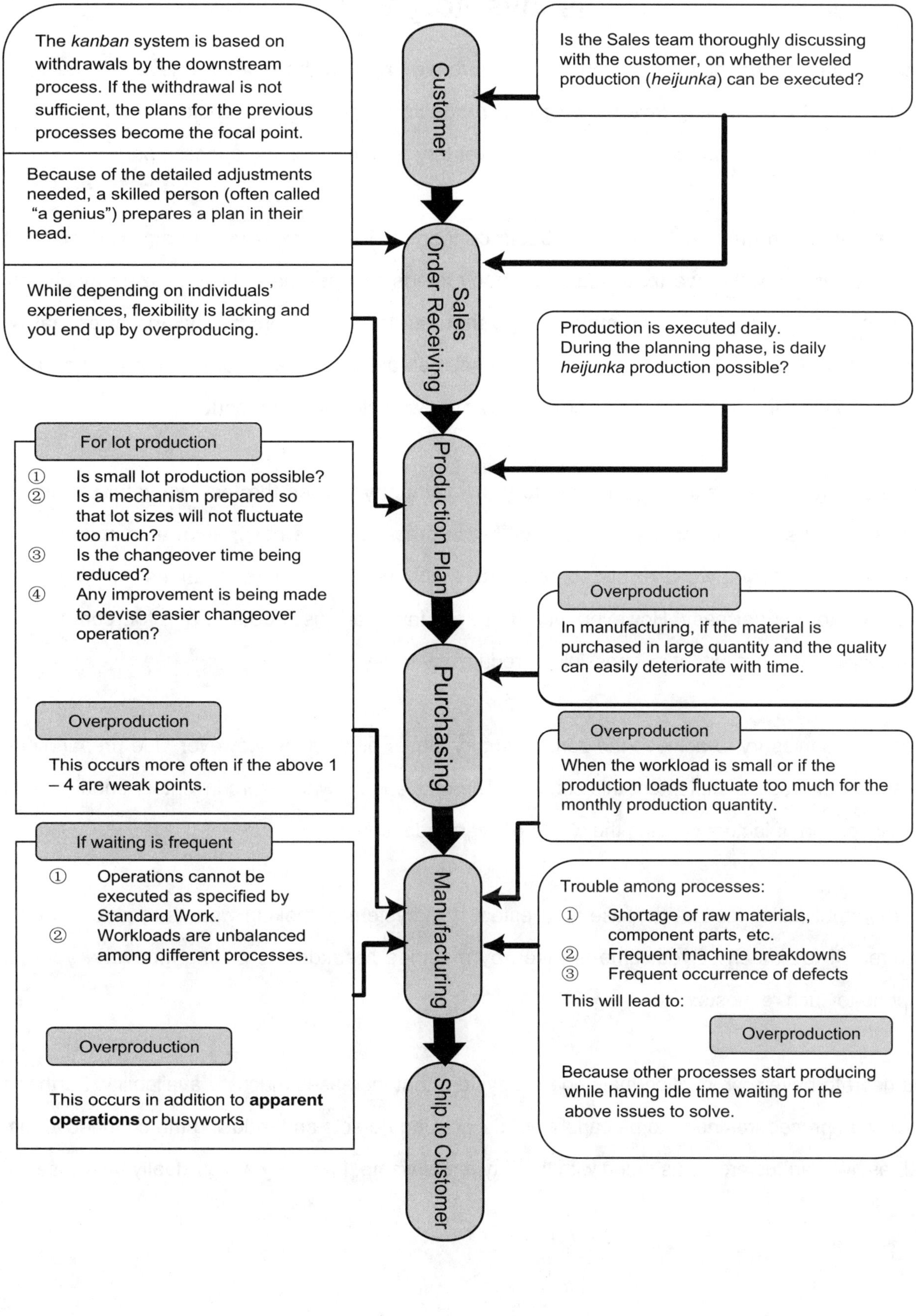

[Process flow: Customer → Sales Order Receiving → Production Plan → Purchasing → Manufacturing → Ship to Customer]

[Pointing to Sales Order Receiving]

The *kanban* system is based on withdrawals by the downstream process. If the withdrawal is not sufficient, the plans for the previous processes become the focal point.

Because of the detailed adjustments needed, a skilled person (often called "a genius") prepares a plan in their head.

While depending on individuals' experiences, flexibility is lacking and you end up by overproducing.

[Pointing to Customer]

Is the Sales team thoroughly discussing with the customer, on whether leveled production (*heijunka*) can be executed?

[Pointing to Production Plan]

Production is executed daily. During the planning phase, is daily *heijunka* production possible?

For lot production

① Is small lot production possible?
② Is a mechanism prepared so that lot sizes will not fluctuate too much?
③ Is the changeover time being reduced?
④ Any improvement is being made to devise easier changeover operation?

Overproduction

This occurs more often if the above 1 – 4 are weak points.

Overproduction

In manufacturing, if the material is purchased in large quantity and the quality can easily deteriorate with time.

Overproduction

When the workload is small or if the production loads fluctuate too much for the monthly production quantity.

If waiting is frequent

① Operations cannot be executed as specified by Standard Work.
② Workloads are unbalanced among different processes.

Overproduction

This occurs in addition to **apparent operations** or busyworks.

Trouble among processes:

① Shortage of raw materials, component parts, etc.
② Frequent machine breakdowns
③ Frequent occurrence of defects

This will lead to:

Overproduction

Because other processes start producing while having idle time waiting for the above issues to solve.

Lesson 9:
Is Inventory Evil?

In this age of abundance with products, the more we produce the more we have leftover products unsold, thereby increasing inventory. Unsold products create various problems for the company's profit. Extra cost will incurred in order to keep inventory, as shown on the opposite page.

In the Toyota Production System the basic concept is to "**make what is required at the time required.**" It is not that we try to make as many things as possible by fully utilizing machines and company resources. Instead, **we make things that can be sold immediately, and after that we do not produce** (even if the machine and other resources are available for production for the day). The point is we think the profit can be generated only after we sell what we produce.

Therefore, it can be said the less inventory the company carries the less wasteful expenses the company incurs. Zero inventory is a concept but in practice I have never seen a company with zero inventory. Inventory always exists somewhere, including at suppliers locations. To operate with zero inventory is not easy. However, with that zero inventory as a goal to aim for, companies ask themselves, "To what extent is it possible to reduce inventory?"

Some companies try to achieve the zero inventory status all at once. However, due preparations are necessary in order to implement this concept. The challenge towards zero inventory should start from the elimination of factors causing the waste of overproduction.

An important factor for that purpose is to establish a system of making what is required at the time required. If too much of your time is taken by machine breakdowns or changeovers, you cannot respond to such requests.

You need to have a production maintenance system that increases machine availability. Furthermore, production procedures need to be capable of decreasing defects and reducing the changeover time. If you resolve various problems found with the 7 types of waste, inventory will gradually decrease.

Figure 9: Problems with and the Purposes of Inventory

Problems with Inventory

(1) The cost to maintain the inventory is incurred

The maintenance cost for inventory is an expense that incurs while maintaining the average inventory of the year, and includes the following:

1. Interest
2. Insurance
3. Transportation cost
4. Warehouse cost
5. Inventory depletion expense
6. Inventory ob solescence expense
7. Tax
8. Facility expense
9. Other expenses

> These expenses are shown as percentage (%) of the annual average inventory value.
>
> $$\text{Inventory maintenance cost ratio (\%)} = \frac{\text{(Cost that incurs by keeping the inventory)}}{\text{(Inventory value)}} \times 100$$

(2) Management becomes weak because the following are hidden:
1. Overlooking defects
2. Carelessness about machine breakdowns
3. Negligence to the need for changeover time reduction
4. Discrepancy between the inventory and the database
5. Maintaining surplus manpower

■ Because of such a company's predisposition, the above problems are hidden in the form of inventory.

Zero inventory
- Implement only with knowledge about the purpose of inventory and the current operational condition of the company.
- To take proper steps for *kaizen* (continuous improvement).

Purposes of Inventory

1. Responding to customer demand fluctuation by keeping extra inventory.
2. Dealing with various problems in production, such as defects, machine breakdown, with the cushion of inventory.
3. Avoiding actions to reduce the lead time, through the use of raw material, work in process and finished goods inventory.
4. Focusing on the cost reduction through mass production (bulk buying, bulk transport).

Lesson 10:
Why Do Companies Not Succeed at Zero Inventory Operation?

The more inventory we have the easier our production will be. On the other hand, in order to aim at having a zero inventory operation the production department of an organization needs to have the following prerequisite in their culture and predisposition.

1) Defects do not occur frequently.

If defects occur often, it becomes difficult to prepare the production plan, and for fear of frequent occurrences, people tend to have a lot of inventories.

2) Machines do not break down often.

If we have machine breakdowns too often, we cannot produce as scheduled, which will cause delays in delivery and troubles to customers. To avoid that, it becomes necessary to have finished goods/ work in process inventories.

3) There is no discrepancy between the inventory ledger and the actual inventory.

There are companies with big discrepancies between the quantities in the ledger and the actual inventories. In such companies, when the production is planned including the inventory quantities, people tend to acquire the habit of having extra inventories because there is a risk that they might experience a shortage at the time of shipment to customers.

4) Various inventory control techniques can be used.

The inventory can be minimized, with the understanding of *kanban* and various other basic inventory control concepts and techniques, such as EOQ (economic order quantity), fixed quantity ordering system, fixed frequency ordering system, ABC control, safety stock quantity, and the like.

On the other hand, if companies without these prerequisites try to achieve a zero inventory operation at once, the production floor will be in confusion. Thus, even if the zero inventory is ideal, there are various issues before achieving it. Because of that, many companies cannot do it until they satisfy various conditions.

Figure 10: Key Points for Implementing Zero Inventory Concepts

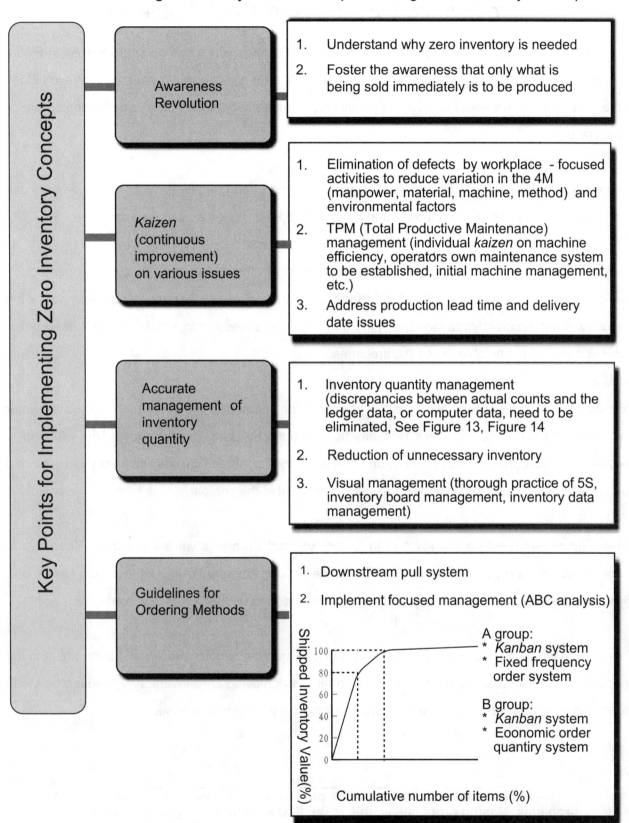

Lesson 11:
Zero Inventory Exposes Internal Problems

The idea is that reduced inventory will expose various problems which can be improved. As in Figure 11 on the opposite page, when there is water (inventory) in the pool (warehouse) problems are hidden in the water and overlooked. However, if there is less water as in Figure 11 previously hidden problems will surface.

For instance, if we stop production when the planned production quantity is done for the day, we can see the appropriate levels of manpower and production speed as well as surplus manpower.

However, operators tend to hate stopping the production in the middle of a production operation. If they continue making products without reducing the production speed, extra products are made. If the inventory is not accurately managed, surplus manpower or excessive inventory will be hidden in the water and we will not be able to see the problems.

Defects and machine breakdowns are big problems for companies but if we have too much water (inventory) we can cope with such problematic situations by using the inventory. This will lead to negligence in recognizing problems not only in the various costs for keeping inventory but also that such a dangerous operational condition has taken root within the company.

On the other hand, with a low level of inventory various problems will be exposed. Then, when problems occur people cannot escape from the situation any longer by using the inventory like before. If you really wish to reduce inventory various problems need to be thoroughly resolved.

Therefore, if you tackle such issues over the years and months by doing *kaizen* and strengthening the operational conditions, it will be a good thing to aim at achieving zero inventory. However, by trying thoughtlessly you might rather have more trouble from the problems that arise.

Inventory hides problems, and the pursuit of zero inventory exposes problems. Making problems visible is the first step in making improvements.

Figure 11: The Role of Inventory

The idea about inventory is as follows:

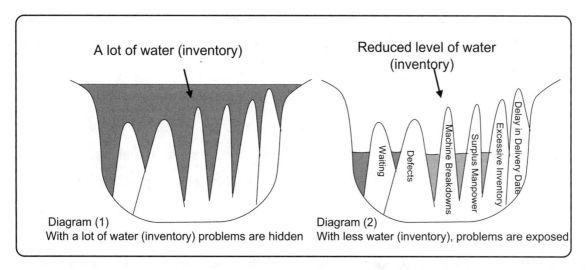

Diagram (1) With a lot of water (inventory) problems are hidden

Diagram (2) With less water (inventory), problems are exposed

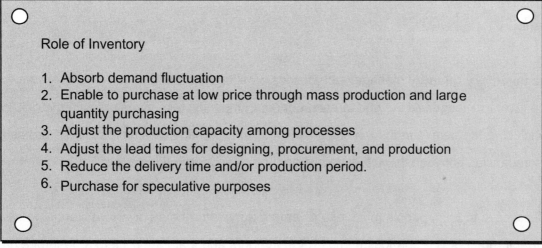

Role of Inventory

1. Absorb demand fluctuation
2. Enable to purchase at low price through mass production and large quantity purchasing
3. Adjust the production capacity among processes
4. Adjust the lead times for designing, procurement, and production
5. Reduce the delivery time and/or production period.
6. Purchase for speculative purposes

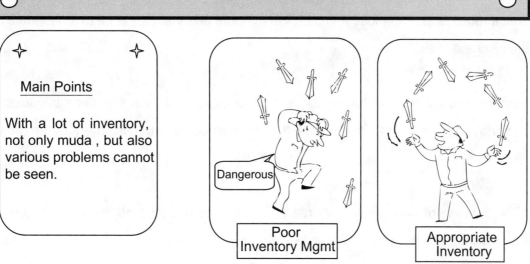

Main Points

With a lot of inventory, not only muda, but also various problems cannot be seen.

Poor Inventory Mgmt

Appropriate Inventory

Lesson 12:
Zero Inventory Begins by Improving Inventory Accuracy

As explained in Lesson 10 as to why companies do not succeed in zero inventory operation, if defects or machine breakdown occur too frequently it is not possible to solve these issues in a short period of time.

If a *kanban* system is to be introduced, the manufacturing operation needs to be structured to do 1) *heijunka* production, 2) changeover time reduction, 3) increase machine availability, and 4) small lot production. Inventory reduction depends on the plans that enable *heijunka* production and meticulous management.

These issues need to be addressed one after the other within the company. Each one of these issues requires the attention and considerable effort not only from inventory-related sections but from all departments.

Accurate inventory quantity management focuses on situations when actual inventory counts do not match the inventory ledger. At the end of each period, companies take inventory with the help of many people, including the people in the materials section. Further, many companies indeed are concerned about discrepancies between the actual counts and the quantities in the ledgers.

If the difference is too big, people must take inventory again after hours or on weekends. In reality it is not rare that it requires an enormous number of man hours and that they fall into a quagmire unable to resolve the discrepancies.

People then acquire the habit of having extra inventories to cope with situations because they feel uneasy that they might not be able to deliver the required quantity to customers by the delivery due dates.

Knowing the purpose of taking inventory (as shown on the opposite page), you will see what can be done by inventory related sections alone. You could tackle this as the first step towards "zero inventory."

Figure 12: The Purpose for Taking Inventory

Major Objectives of Taking Inventory

(1) Confirm the quantity on the inventory ledger and actual goods, to know the differences. If there is no discrepancy, production can be arranged and accurately managed. Any discrepancy needs to be investigated to determine the cause, while taking measures to confirm accurate numbers.

(2) In the process of confirming actual goods, you might encounter dead stocks that will be unlikely to be used, or inventory that is gradually deteriorating in quality. You then would try to make effective use of such inventories.

Additional Reasons for Taking Inventory

(1) Confirm the storage condition of the actual goods
(2) Verify whether the stock quantities are appropriate:
   a) Confirm slow moving materials, and take countermeasures
   b) Reduction of inventory carrying cost
   c) Check if the composition of raw material inventory is adequate
(3) Study the storage methods and paperwork processes:
   a) Efficient delivery methods, in and out of warehouses, can be seen
   b) Identification management can become thorough, for deliveries in and out of warehouses, including returns, dead stocks
(4) Inventory management structure becomes strong, if the inventory ledger matches the counts of actual goods
   a) If the discrepancy is large, people cannot rely on the quantity in the ledger and tend to order larger quantities of materials than needed
   b) Thinking that a small discrepancy cannot be verified, it can result in the dishonesty of employees

Lesson 13:
To Improve Inventory Taking Accuracy

Inventory taking is conducted so that we confirm the actual quantity of goods and the ledger data, as explained earlier. However, many companies have a hard time with unreconciled discrepancies between the inventory counts and the inventory ledger data.

People often feel despair at such results of inventory taking with many discrepancies, after spending so much of manpower and time. However, companies with such results will find discrepancies even if they take inventory many times. Inventory taking will never be effective unless the root causes are investigated and changed in order to improve the company's operational fitness.

Such companies would find it useful if they focused on products with lots of discrepancies and **investigated** them for **true causes of deviation** during a certain period of time. At Toyota they practice the concept of *jidoka* (autonomation or automation with human intelligence) in which they stop the machines when problems occur and investigate issues without waiting.

Inventory discrepancy is the same. By investigating the causes without waiting, the true causes can be found and easily resolved. If the true causes become known, quite often they can be resolved even without advanced countermeasures.

The diagram on the opposite page is a part of a sample case where thorough investigation was conducted on the parts with many discrepancies between July 15 and July 31 for the condition of deliveries in and out of warehouse. Every day the actual counts of goods and the ledger data in the computer were verified and the reasons for the discrepancy were searched immediately without leaving them unresolved.

It requires much effort as it entailed the comparison of the actual counts of goods every day with the data in the computer. However, by taking the time to investigate the issues which cause the discrepancies, the weak points within the operations of the company can be identified.

> **Jidoka is the ability of machines or processes to detect problems and stop autonomously. This is also called "automation with human intelligence".**

> Case example

Figure 13: Investigation on Deliveries in and Out of Warehouse by Distribution Control Department

1. Our Sales Department placed the order with Sunny Industry (manufacturer), who delivered 100 pieces to our Distribution Management Department.
   a) Provisional Delivery Voucher was attached to the product boxes, delivered by Sunny Industry.
   b) These were accepted as sample items before mass production.
   c) Despite the data given by Sales Department, our purchasing staff has not issued order form because of the nature of sample items.
   d) Therefore, Mr. Sugimura of Distribution Mgmt Dept requested Sales Dept dispatch the Special Voucher for Delivery into Warehouse 7/15 Mr. X; FAX Mr. Y → Mr. Z.

2. After inspection, 83 pieces were considered good products and delivered to AH Company.

3. The Provisional Delivery Voucher was not processed for storing in the warehouse. However, its sales transaction portion only was processed in the computer. This created negative balance of the item with minus 83 pieces.

Investigation Sheet for Parts - Storing in and out of warehouse

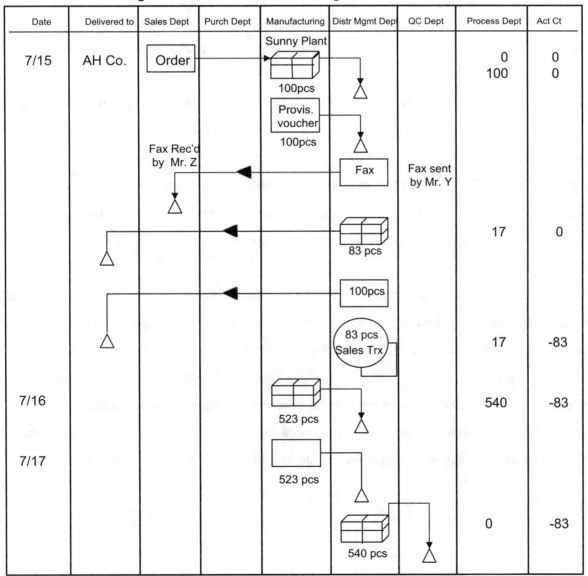

Lesson 14:
Case Example of Inventory Taking

Various problems of inventory are indeed condensed in the case example of Lesson 13. The remaining thing is to establish the standard for actual goods, paperwork, and the communication for these so that such problems should not occur. Companies with lots of problems might be better off if they try investing the issues in this way.

If properly done at some point the discrepancy will be resolved between the actual goods and the ledger balance in the computer. As in the example, if only the items approved by inspection are delivered out of the warehouse because 1) the sample status before mass production, 2) the data was processed with provisional voucher or 3) the quality standard was not clear for the delivery into the warehouse, the information lags more and more behind the actual goods.

Also, as more departments are involved with the information (paperwork, etc.) and the movement of the actual goods, greater discrepancies can be caused between the information and the actual goods.

In the case example, 523 pieces of product were delivered into the warehouse on July 16 together with a provisional packing slip (because the approval was not obtained for the 83 pieces of products delivered to AH Company on July 15th) there was a discrepancy of 623 pieces on July 17.

Then on July 23 the information caught up with the actual move of goods and the discrepancy disappeared. However, the discrepancy started to appear that afternoon again. With such inventory inaccuracy, the zero inventory operation is out of the question.

Issues exposed by this case example need to be resolved. Companies with inventory discrepancies who do not know the causes of such problems might try to investigate the problems using such an Inventory Discrepancy Tracking Survey Sheet as shown in figure 14 on the opposite page.

Figure 14: Inventory Issues Exposed in Case Example

a) No rule has been formulated for the delivery in and out of the warehouse for prototype models, using special vouchers.
b) For mass production items in circulation, vouchers are processed after the physical transportation of products, and therefore the inventory balance is not captured real time in the computer. In other words, the actual goods are not moving together with vouchers.
c) There are too many transactions using provisional vouchers. Many transactions exist using the provisional vouchers in the following route: Manufacturer -> Distribution Management Dept -> Quality Control Dept -> Processing Dept. Because of this, discrepancies between the actual goods and computer balances occur, which makes it impossible to see the real time status of actual goods.
d) It is not clearly defined as to who within the originating section should handle 1) the engineering prototype making, 2) the trial production before mass production phase, and 3) the mass production items. The responsible section within the receiving party needs also to be clearly defined.
e) Sometimes products are left loaded in the truck. Because of this, it is not possible to make decisions on discrepancies between the actual goods and the computer balance (ledger inventory) for the inventory in transition.
f) The procedure for processing vouchers for when Quality Control Department requests other sections to process some modifications is not clearly defined.
g) Warehousing processes for Distribution Management Dept and Quality Control Dept are not conducted real time.

Figure 14.1: Discrepancy Tracking Sheet

										Doc No.
		Products Name				Products No.	Date Prep'd	Approval	Inspection	Preparer
Date	Vcher Y/N	Quantity Entered Entered	Vcher Y/N	Quantity Delivered Out of Warehouse	Inventory Balance (Actual)	Inventory Balance in Leder	Inventory Balance in Computer	Reason for Discrepancy		Remarks

Lesson 15:
The Overall Picture of TPS

The Toyota Production System (TPS) is built on just in time and *jidoka* (autonomation) as its two pillars. Just in time means to "make only what is required, at the time required, in the quantity required."

In concrete terms the downstream process goes to the upstream process and from there withdraws only what is needed, when it is needed, at the quantity needed. In this way the upstream process makes only the quantity consumed in the downstream process, and creates no unnecessary inventory. This system is based on its prerequisite, *heijunka* production.

Jidoka (autonomation) is sometimes called "automation with human intelligence" and means a process with machines in which human intelligence has been installed. If any flaw appears the machine automatically monitors and controls it.

A concrete example is that when a defect occurs, the machine with an automatic stop mechanism will stop the machine operation to avoid the flow of defective products from the upstream process to the downstream processes, and from causing problems in the following work centers.

On the other hand, the "automation **without** human intelligence" continues making defective products even after the abnormality surfaces and leads to massive defects and machine trouble.

The habit of stopping the machine immediately when flaws appear and finding the causes is results in the elimination of defects based on "go see" mind set. This entails looking at the defects and observing the condition of defect occurrence, making the problems easier to fix. In this approach there is no need for people to be always stationed at machines and *kaizen* will progress and **flexible manpower** becomes possible.

The TPS house with the two pillars is illustrated on the opposite page. Just as anyone without proper training cannot climb up to the summit of tall mountains, the development of TPS at Toyota has been the result of organizational efforts over many years. They developed people who can do *kaizen* and they have now the culture of eliminating even the tiniest waste.

> **"Make it blossom with originality and creativity."**
> **Sakichi Toyoda (Founder of Toyoda Loom Works)**

Figure 15: Outline of Toyota Production System

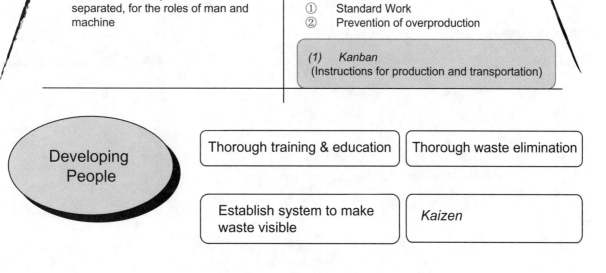

High Quality, High Profit

Toyota Production System

Jidoka (Autonomation)

Quality is built in at the process

① Abnormality stops operation
② Abnormality is visible (Visual Control)
③ Problems solutions is based on gemba principle (*genchi, genbutsu, genjitsu*) and workplace-focused action

Flexible manpower

Roles to be clearly defined and separated, for the roles of man and machine

Just in time

Flows in the process

① One piece flow
② (Small lot Production)
③ Changeover time reduction
④ Synchronization
⑤ Multi-process handling
⑥ Multi-skilled workers
⑦ Machine layout in synch with
⑧ sequential process

Takt Time

① Standard Work
② Prevention of overproduction

(1) *Kanban*
(Instructions for production and transportation)

Developing People

Thorough training & education

Thorough waste elimination

Establish system to make waste visible

Kaizen

Section 2:

Just In Time

Lesson 16:
What is Just In Time?

Just in time means each process receives "what is needed, when needed, in the quantity needed."

Manufacturing plants make efforts to always produce according to plans and ship the products out according to the delivery due dates, but deliveries of component parts too early to the warehouse create the waste of inventory. Deliveries that are too late cause delays in shipping to customers. Thus, the reality is not easy.

Just in time has the prerequisite of *heijunka* (production smoothing) and consists of three concepts, namely making the process flow, determining the takt time based on the required quantity and withdrawal (pull) of materials by the downstream process.

Making the proceses flow means that by practicing one-piece flow in processing or assembling, the flow improves. For industries with machines and equipment there is a need for small lot production as much as possible, therefore efforts in reducing setup times are needed.

Factors such as equipment layout in sequence of the processes, multi-skilled operators handling a number of processes, and training for multi-skilled operators will reduce the stagnation or lack of flow and make the flow smooth. Then synchronization through standard work becomes necessary, and the takt time will be determined. With this takt time it becomes possible to avoid overproduction.

Withdrawal (pull) of materials by the downstream process means that the upstream processes make only what has been consumed by the downstream processes. Behind the downstream processes are ultimately the customers as the final process. In other words, it is a system of making only what the customers ordered.

As a means of operation of downstream pull, production *kanbans* and withdrawal *kanbans* will be used. Just in time looks at the production flow, from the opposite direction (from the downstream processes).

> **"It is important that each component part be ready just in time."**
> **Kiichiro Toyoda (President of Toyota Motors, 1941 – 1950)**

Figure 16: The Purpose of Just In Time

Purpose	(1) Flexibly handle demand changes (2) Eliminate waste of overproduction (3) Reduce lead time

Figure 16.1 How Just In Time is Structured

Basic Idea	Actions	Examples
Making processes flow	(1) Draw ideal flow lines (2) Synchronization of material consumption (3) Multi-functional operators, multi-tasked operators for a number of processes, stand-up work (4) Machine layout in the sequence of processes (layout along the flow)	(1) Use of work design (2) One piece flow, small lot production for industries with equipment (3) Synchronization among processes (U shape line, parallel line) (4) Machines and equipment are placed in sequence of the process flow
Determining the takt time in the line, based on the required quantity, in response to the changing demand	(1) Through practice with Standard work (takt time operation sequence, standard WIP inventory) (2) Mechanism alarming abnormalities in the line	(1) Cross training board, Standard Work Combination Sheet, Standard Work Sheet, Work Instruction Sheet (2) Automatic stopping mechanism (3) Operators wait when idle time between processes occurs
The downstream processes withdraws from upstream processes	(1) *Kanban* (2) Transport without using kanban	(1) Production *kanban*, Withdrawal *kanban* (2) Fixed quantity fixed time transport, fixed time variable quantity transport, variable time fixed quantity transport, variable time variable quantity transport
Small lot production	Changeover Time Reduction	SMED (Single Minute Exchange of Dies)

Lesson 17:
Heijunka as a Prerequisite of Just In Time

A prerequisite of Just In Time is *heijunka* (production smoothing). *Heijunka* stabilizes the flow of production and avoids unevenness and variability (*mura*).

Ordinarily the difference between the **man-hours** (based on the production requirement) and the production capacity is investigated at the plant, **using the cumulative workload method.** If there are imbalances in the workload (balance between the **man-hour** requirements and production capacity) for each process and for each machine, unevenness in performance will occur, which will result in waste. To minimize unevenness is also *heijunka*.

Let us apply this concept to a "high mix, low volume" production. In the lot production of figure 17.1, if the production in the downstream process (assembling) fluctuates, the upstream process (making product A) is busy at the beginning but from around the time the downstream process starts to make Product B, we start to see that the upstream process is underutilizing the workers' time.

During a busy period, people cope with this situation by having extra resources (machines, manpower, etc.) and extra large inventories, in accord with the upswing of the workload. When *heijunka* is practiced as in figure 17.1 by removing the large fluctuations in the downstream processes, the workload at the upstream processes can be lowered, and thus, stable production becomes possible everyday.

In order to enable balanced production, not only the averaging of volume, but also that of item variety (mix) will become necessary. At Toyota, such a balancing operation of volume and product mix is called *heijunka* (production smoothing).

However, it is not easy to cope with customer orders which vary in mix and volume. *Heijunka* production is good as concept but in order to implement it various requisites need to be completed before that. Inventive efforts to increase the fitness of the organization are necessary in order to succeed with *heijunka*.

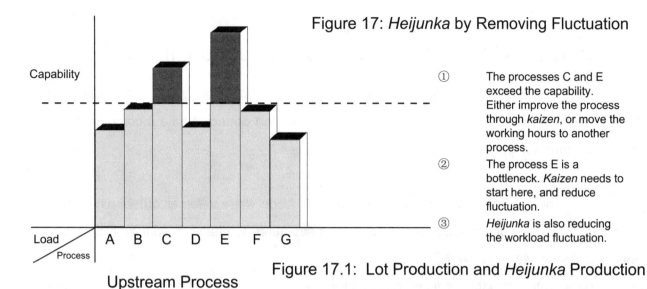

Figure 17: *Heijunka* by Removing Fluctuation

① The processes C and E exceed the capability. Either improve the process through *kaizen*, or move the working hours to another process.

② The process E is a bottleneck. *Kaizen* needs to start here, and reduce fluctuation.

③ *Heijunka* is also reducing the workload fluctuation.

Figure 17.1: Lot Production and *Heijunka* Production

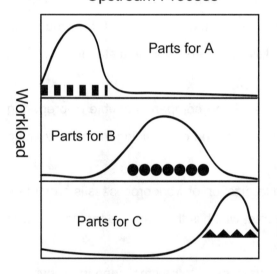

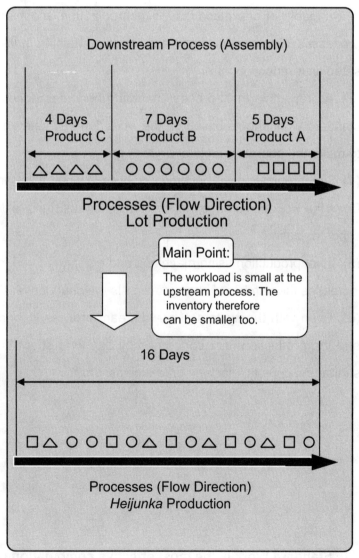

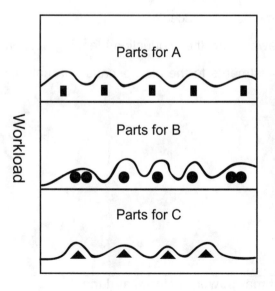

37

Lesson 18:
Why Do Companies Not Succeed at Just In Time?

Just in time is to receive supply for "only what is needed, when it is needed", in the quantities needed" at each process, it may create confusion instead within a short period of time if companies cannot make improvements in the following areas, including *heijunka*.

(1) In a continuously changing market, it is not easy to cope with customer orders which are not fixed in terms of variety and quantity. Is the production structured to enable *heijunka* which levels out quantities and varieties every day?

(2) Is the organization capable of flow production in the manufacturing processes? In other words, is it capable of one piece flow, synchronization, machine layout in the sequence of the production processes, repetitive multi-process handling, training multi-skilled workers, work in a standing position, setup time reduction, etc?

(3) As repetitive multi-process handling becomes necessary, is the company capable of preparing standard work sheets based on the work done by veteran workers, and of providing thorough training to make the operators multi-skilled?

(4) Synchronization will become established when the standardization of work processes is promoted. Does the work place have the structure of establishing takt time, which is the standard for the speed of synchronization?

(5) In lot production using machines and equipment, the lot sizes need to be small, and this inevitably increases the number of changeovers. Is the changeover time being reduced?

(6) Even with the system of downstream process withdrawals, are the production planning section and the production floor concerned all day long about adjustments to the plan because of material shortages, defects, machine breakdowns, etc?

> **Standard Work is the most effective combination of manpower, material and machines based on takt time, work sequence andstandard work in process stock quantities.**

Figure 18: Main Points in *Heijunka* (Production Smoothing)

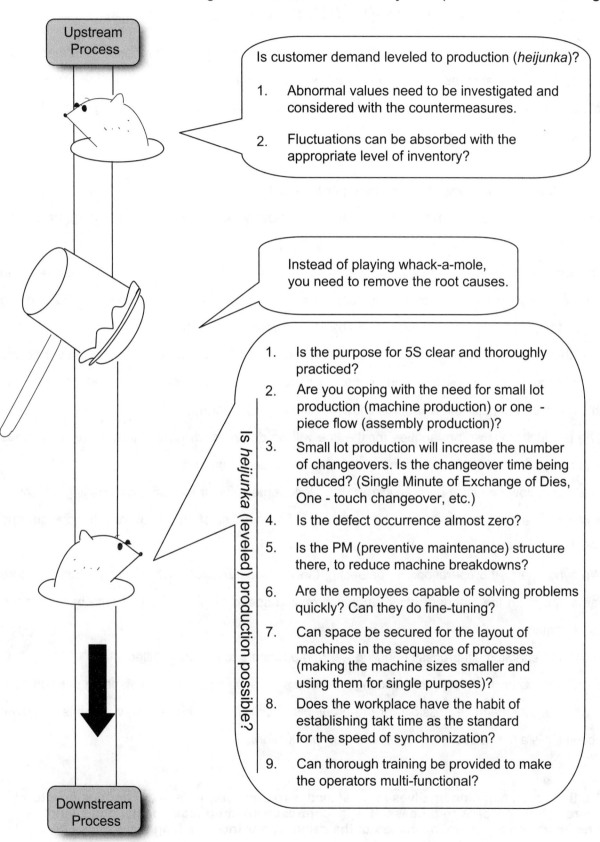

Lesson 19:
Making Processes Flow

Flow production is a system where you change the traditional idea of work proceeding as processes, and think of the processes as a "flowing stream" where you look at the products as the stream. The ultimate goal of making the process flow is one-piece flow production by removing stagnation of things within and between the processes.

To achieve flow, the following will be the prerequisite conditions:

(1) In order to make flow, we need to visualize ideal flow or stream (Refer to the diagram on the opposite page).

(2) Arrange machines in the sequence of the processes, and remove wasted transportation. Also, in principle we use small machines and equipment that are single-purposes (as opposed to multi-purpose use) so that the machines and equipment can be easily integrated into the line.

(3) The starting point of unprocessed items and the exit point of finished items need to be very close to each other as much as possible, and in order to eliminate waste of motion we use a layout such as a u-shape line or a parallel line layout (as shown on the opposite page).

(4) The product types and quantities for the line will be determined. Before that however, various obstacle factors for *heijunka* production need to be tackled and improved.

(5) One-piece flow needs to be practiced at the work centers for processing, assembling, finishing, and small lot operation for production with machines. The number of changeovers will increase, and we need to reduce the changeover times.

(6) We synchronize the operations by balancing the workload so that each process will have about the same speed. In the standard work the synchronized speed must be the takt time based on the customer demand.

(7) In order to enable multi-process handling we train operators to be multi-skilled.

(8) In order to enable multi-process handling it becomes necessary to work in a standing position.

(9) Prompt actions become possible for the problems arising from one-piece flow and this improves production skills and realizes a higher level of flow production.

> **Multi-process handling involves multi-skilled workers working in a continuous flow to progress from process to process. It is in contrast with multi-machine handling which is the operation of several machines of the same type without making the process flow.**

Figure 19: The Ideal Flow

The ideal flow concept

• At Toyota there is a basic idea that any activities that do not produce added value are waste. That is illustrated specifically as 7 types of waste.

• All the waste will lead to the waste of waiting. Thus, if we eliminate the waste of waiting we can be closer to just in time.

Actions to achieve the ideal

• For that purpose, obstacles to the flow need to be thoroughly investigated and we must visualize the ideal flow or stream of when these obstacles have been removed.

• However, the ideal flows imagined by companies with many problems are not realistic. The most simple and effective system and method needs to be visualized, in such a way that the objectives can be achieved.

Figure 19.1: Transforming Muddy Streams into Clear Streams

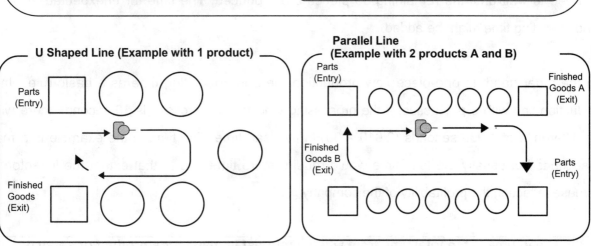

Lesson 20:
What is Production Lead Time?

The definition of production lead time varies slightly from company to company, but at Toyota it is "the time elapsed from the time we start processing materials into products to the time we have finished goods. It is the total of the processing time (the time to increase the value added) and waiting time (the time that does not add value)."

Lead time can have various patterns, but in general these can be classified into the build to forecast and the build to order. By shortening the production lead time, the supply to customers can be faster which can lead to greater customer satisfaction. Within a company, it can help reduce risk and increase the adaptability to the changes in environments and circumstances.

Lead time is related to lot size. The bigger the lot size is, the longer the lead time becomes and the larger inventory as well.

For instance, as shown in the diagram on the opposite page if the same thing is produced in a batch the production lead time is very different from when one piece is made at a time. In this example the additional times for transportation, waiting, etc. have been omitted for the sake of easier understanding.

In **lot production**, the time to make one piece is the same, but as the production is in a batch mode we need to wait until the remaining 99 pieces are produced. The time for unexpected events, or various waiting time might be added.

On the other hand, in one-piece flow production there is no waiting time as a basic rule. In the production line with 3 processes, if the processing time for one process is 10 seconds there will be the difference of 1,980 seconds (2,010 – 30) in the lead time. The time in the example is a matter of seconds but usually the lead time would require more time and to that extent the inventory will increase, affecting the profitability of the company.

> **Lead time is the total time from when a request is made for a product or service until that request is fulfilled, and includes all processing time as well as waiting time.**

Figure 20: Production Lead Time

Two general classification of production lead time

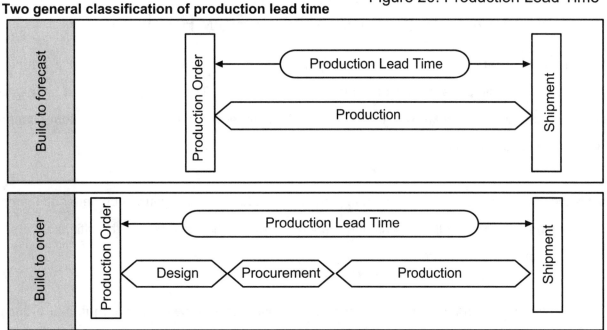

Figure 20.1 Lead Time for Lot Production vs. One Piece Flow

Difference between Lot Production and One Piece Flow Production

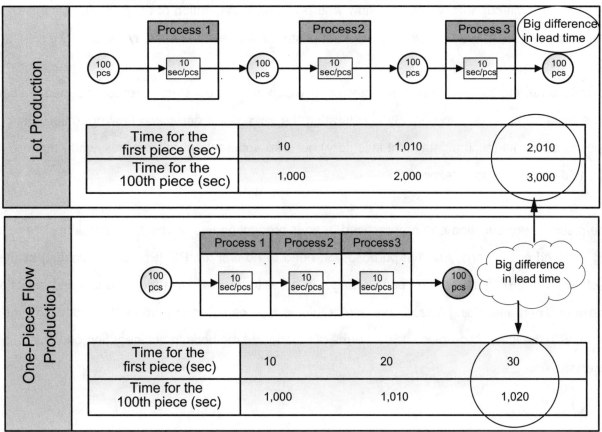

Lesson 21:
The Meaning of One-Piece Flow Production

Lot production creates a lot of waste in inventory and man-hours used to meet production requirements. Is it simply a matter of implementing *heijunka*? The reality is not that simple, as companies attempting *heijunka* without practicing the changeover time reduction for item changes would experience complications.

If that is the case, instead of focusing on one-piece flow production alone you will be better off thinking of the meaning of one-piece flow. By nature, the just in time is characterized by the ultimate fact that waste can be exposed anywhere.

One-piece flow is a method of processing or assembling one piece only in sequential order of processes. By practicing one-piece flow production, problems between the processes can be exposed. At Toyota the flow of things between the processes is viewed as if it is a stream of water. Narrow streams with swift currents will clearly show where you have stagnation or things accumulating. If you remove the causes of the problems, the flow of the stream will return to the way it was.

Operations are the same, and with one-piece flow which is a narrow stream, issues that had not been detected before can be exposed. By accelerating the speed of a one-piece flow operation, not only exposed problems (such as stagnant inventory) but also apparent operations (busywork) that are not easy to detect can also be exposed.

One-piece flow production can be practiced in your production line, when such problems are found and resolved one by one. For that purpose you need to do first the PQ (Products Quantity) analysis as shown in the diagram on the opposite page, in order to narrow down the products and select the problems. Then after doing *kaizen* on these problems, you can start to prepare the production line in accord with the idea of one-piece flow in terms of manpower, methods, machinery and equipment, and materials.

Figure 21: The Meaning of One-Piece Flow

Principal Viewpoints of One-Piece Flow Production

PQ Analysis

P (Products)
Q (Quantity)

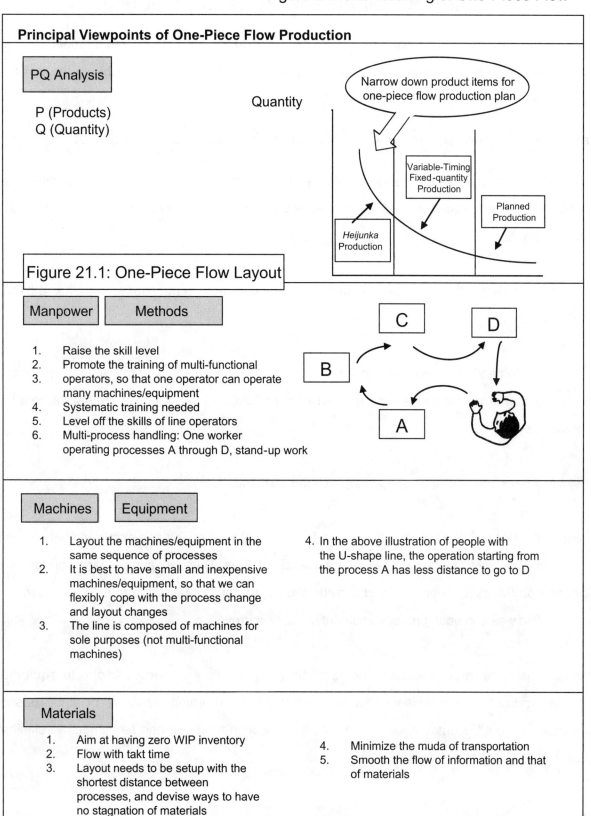

Narrow down product items for one-piece flow production plan

Quantity

Heijunka Production

Variable-Timing Fixed-quantity Production

Planned Production

Figure 21.1: One-Piece Flow Layout

Manpower   **Methods**

1. Raise the skill level
2. Promote the training of multi-functional
3. operators, so that one operator can operate many machines/equipment
4. Systematic training needed
5. Level off the skills of line operators
6. Multi-process handling: One worker operating processes A through D, stand-up work

Machines   **Equipment**

1. Layout the machines/equipment in the same sequence of processes
2. It is best to have small and inexpensive machines/equipment, so that we can flexibly cope with the process change and layout changes
3. The line is composed of machines for sole purposes (not multi-functional machines)

4. In the above illustration of people with the U-shape line, the operation starting from the process A has less distance to go to D

Materials

1. Aim at having zero WIP inventory
2. Flow with takt time
3. Layout needs to be setup with the shortest distance between processes, and devise ways to have no stagnation of materials
4. Minimize the muda of transportation
5. Smooth the flow of information and that of materials

Lesson 22:
What is the Required Takt Time for Production?

Takt time is the time to produce one piece of product and represents the operation speed. Efficient production means that the processing time of each process is as close as possible to the takt time. Takt time is calculated as follows:

> Takt time = Available time for production per day / Required production quantity per day

Takt time is based on the quantity needed by the customer. Referring to the data on the opposite page, if the customer demand for the product is 400 pieces a day the takt time becomes 49.9 seconds. More production than that will simply increase the inventory.

Efficiency will increase if the same thing is made continuously, but we may suffer from the setback of making too much.

If the customer demand decreases, say to 200 pieces a day, then we should either stop the production and do something else or decrease the number of operators for each process so that the takt time becomes longer.

Avoid slowing down the operation speed to match the takt time.

After setting the takt time the next thing to do is to decide operation speed and workload sharing based on the current capability. Operators have different levels of skills. Thus, we can prepare standard work sheet based on the skilled operator and provide thorough training based on that. Also, it becomes important to find waste in each process and continuously improve.

Address all lost time through *kaizen*, including time lost within each process due to equipment breakdowns and changeovers, as well as time lost to breaks, morning meetings, or other reasons. For the defect ratio it becomes necessary to study inspection methods and techniques to eliminate defects.

Figure 22: Required Takt Time for Production

1) Takt time is determined by customer demand

 Suppose customer demand for the product is 400 pieces a day

 Calculate the available time for production, in a shift of 8-hour operation

Actual operation time
= (Planned operation time – time outside of the planned operation time) x Operation Ratio
336 min = (8 hr x 60 min – 1 hr x 60 min) x 0.8

 Calculate the production quantity

Production quantity = Customer demand quantity / (1 – Defect ratio)
404 pieces = 400 pieces / (1 – 0.01)

 Calculate the takt time

Takt time = Actual available for production / Production quantity
49.9 sec = 336 min x 60 sec / 404 pieces

2) Explanation of terminology

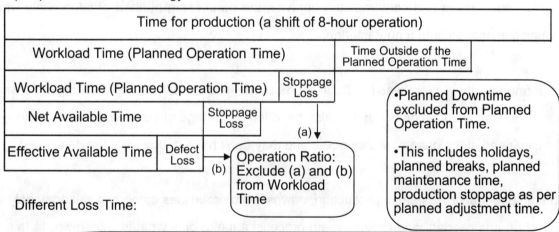

(a) Breakdown loss, changeover and adjustment loss, tool change time loss, ramp-up time loss
(b) Minor stops, reduced speed
(c) Rework time loss, time yield loss, loss due to scrap and defects

3) When customer demand changes to 200 pieces a day
    a) Stop the work
    b) Decrease the number of operators and make the takt time longer

Lesson 23:
Withdrawal (Pull) by the Downstream Processes

The meaning of withdrawal by downstream processes is that the upstream process will make only what is withdrawn by the downstream process. Why do we have this idea? It is because it has been revealed that in the traditional production system (called push system) the decisions by the upstream processes would dictate production, creating enormous waste in inventory, storage, transportation, etc.

In a traditional production system when things are made in the upstream process they are brought to downstream processes.

If the upstream process has extra capacity with machines, equipment and manpower, they would make more and more things. If the quantities produced match the quantities required then it is fine but in many cases it is not the case. Some products need to be stored at warehouses. Then like perishable foods missing the opportunity to be sold, these can become dead stock.

To make things just in time is a good way to avoid creating waste. Taiichi Ohno took a hint from the supermarket sales system and contemplated applying the purchasing method of taking only what is needed when it is needed, into manufacturing.

At the supermarket, they avoid wasting fresh foods purchased in bulk by replenishing only what is purchased by customers. Customers in turn also have the advantage of not having waste, as they buy only what they need at the quantities they need, and they don't have to buy what the don't need.

Adapting the supermarket system in production, downstream processes come and withdraw only what is needed when it is needed, and the upstream processes make only what is withdrawn. In this way, the problem is avoided as people don't have to make what is not necessary.

Figure 23: What is Withdrawal (Pull) by Downstream Processes?

1) Traditional Production System (Push System)

Based on production plans, the materials are purchased and the production environments and resources (manpower, methods, machines and equipment) are prepared and the products are delivered by the delivery due date customer requested.

- When materials are at while the production plan is still at its preliminary preparation stage the purchased items simply become inventory at the time when the plan becomes firm with changes, and if there is no customer order it becomes a dead stock.
- In the production process the information flow goes far ahead of the physical flow of materials and therefore each process produces in accord with plans. Naturally, as you go more and more to the downstream processes things are pushed. Also, in the production process there is a tendency to make more in anticipation of defects, lower yield for not passing inspection, machine breakdown, etc.
- The more changes we have to our plan, the more waste will be created in not only that of product (inventory) but also that of man-hours, including the adjustment of production plans.

(2) Supermarket Sales System (Downstream Pull System)

The flow of information moves by small units and soon after the information, the physical moves of materials take place and therefore people can swiftly cope with the changes in planning or production, which reduces waste.

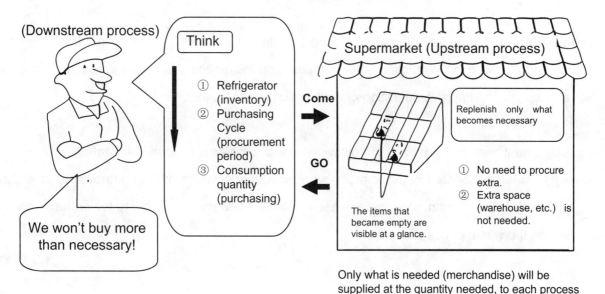

Only what is needed (merchandise) will be supplied at the quantity needed, to each process (merchandise shelf).

Lesson 24:
What is the *Kanban* System?

In just in time production, withdrawal of materials by the downstream pull system is practiced. The control tool used for this purpose is *kanban*.

The *kanban* system is a communication method for transmitting production information using *kanban*, or signals to trigger actions. It is a production control technique representative of TPS that organizes the way to make things and also the flow.

In the traditional push system, the information flow goes ahead and the things made at the upstream processes are pushed to the downstream processes one after another. In the downstream pull system, **the *kanban* and the materials move at approximately the same time.**

The work center assembles only what is sold to customers, and the component part work center processes what has been used by the assembling work center. In this way the operation proceeds sequentially to the upstream processes.

After the demand is originated the information flows without a time lag in this system. Therefore, a flexible response to changes in production plans becomes possible without waste.

As shown on the opposite page, *kanban* has the functions of three types of instruction. They are identification tags, production instruction, and transportation instruction. Also, the purpose of *kanban* is to achieve better quality, improve operations, and to decrease inventory.

Visual management becomes possible, by following up with problems exposed by *kanban*.
Too many *kanbans* represents too many production instructions, which might contribute to inventory increases. Therefore, by making effort to decrease *kanban*s it becomes possible to lower the inventory levels, avoiding the waste of overproduction.

Visual management is the use of visualization tools to quickly detect and correct abnormalities.

Figure 24: The Three Functions of *Kanban*

(1) Flexibly handle demand changes

(2) Eliminate the waste of overproduction

(3) Reduce lead time

Figure 24.1: The Purpose of *Kanban*

Quality Improvement	• To have a flow of "what is needed, when needed, in the quantities needed," means that no mistakes are allowed, and any mistakes need to be corrected right away.
	→ This matches the traditional (non-mass production) quality control of "the quality built in at the process."
Operational Improvement Tool	• When *kanban* and the material are together, you see the item name, item number, quantity, right away at a glance. Thus, the *kanban* becomes a tool for visual management. When *kanbans* remain in certain processes, or there are no *kanbans* to be seen, it means that the operation is being tied-up somewhere.
	→ By looking at how the *kanban* are kept tied-up on the production floor, the progress of the operation can be captured and the points for improvements become clear at the gemba (actual place of work).
Inventory Reduction	• With the number of *kanban* it becomes easy to grasp the inventory. Too many *kanban* means there are too many inventories.
	→ By trying to reduce *kanbans* it becomes possible to reduce inventory while avoiding waste of overproduction.

Lesson 25:
The Types of *Kanban*

Kanban can be generally classified into two types. One is the production instruction *kanban*, which is used within the processes to give production instruction. The other is the withdrawal *kanban*, which is used to withdraw component parts into the production line.

(1) Production instruction *kanban*

It is also called production (processing) *kanban*.

a. In-process *kanban*

This in-process *kanban* indicates the type and quantity of component parts to be produced in the upstream processes. This type of *kanban* is used to order operation for the processes, in which almost no changeover time exists (such as the production line for one type of item only).

b. Signal *kanban*

When we process multiple types of items in one production line the changeover time inevitably occurs. This type of *kanban* is used for processes with a certain level of lot production.

(2) Withdrawal *kanban*

It is also called transport *kanban*.

c. Supplier *kanban* for outsourced items

This is used to withdraw component parts from outsourced makers, which are the upstream processes for the operation.

d. Withdrawal *kanban* between processes

This is used for withdrawals between processes; to withdraw the component parts for assembly lines.

Written on the *kanban* are: item name, item number, package type, container capacity, line name, number of *kanban* in circulation, upstream process name, downstream process name, order multiple, etc.

For the outsourced component parts (like the diagram on the opposite page) we add other information such as the supplier company name, storage location of supplier company, delivery cycle, recipient company name, receiving location, storage location of recipient company.

Figure 25: The Types of *Kanban*

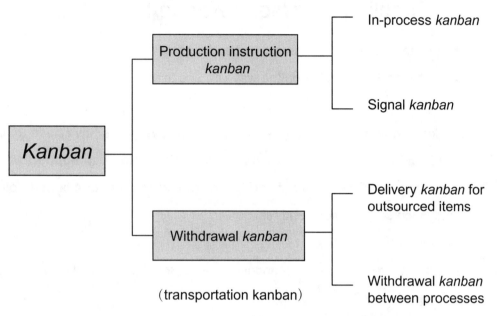

(transportation kanban)

Figure 25.1: *Kanban* Card Example (supplier *kanban*)

Delivery Time	Delivery Location (Storage address, Street, City)		Recipient Name	
Delivery Cycle Bar Code	Bar Code		Receiving # Bar Code	
Supplier Co. Name	Item Name	Number of *Kanban* issued	Receiving location	
Supplier storage location	Item #	Package type		
Post #	ID Tag at the back	Box capacity		

There are many types of *kanban*.

Each company can prepare what is best for them.

Lesson 26:
Enabling the Use of *Kanban*

Kanban is a tool to communicate information on the production flow and the way to make things. Tools can be either useful or rather harmful, depending on the way you use them.

It is important to know that *kanban* **can work only when various improvements have been made to the actual workplaces and that the operational fitness of the organization can support** *kanban*. If companies that lack *heijunka* (production smoothing by averaging of product mix and volume) implement *kanban*, they may face complications and feel busier.

The prerequisites for *kanban* use are shown on the opposite page. Hopefully, these will be useful for you.

Further, the rule for the use of *kanban* is as follows:

(1) Downstream processes withdraw *kanban*

If *kanban* either remain at certain locations or moves ahead of time it means we have some problems. There are three rules for *kanban* withdrawal.

    a. No *kanban*, no withdrawal

    b. No more withdrawal than the quantity specified by *kanban*

    c. *Kanban* must be kept attached to the actual goods (or containers)

(2) Produce only the quantity withdrawn by downstream processes

Minimize the inventory in each process. There are two rules for the production using *kanban*.

    a. No more production that the quantity specified by *kanban*

    b. Produce in the sequence the *kanban* cards arrive

In order to have a *kanban* system successfully implemented, the level of the system needs to match the operational capabilities of the company. Also, you need to set the rules for *kanban* and make thorough efforts in maintaining and following what was agreed.

Figure 26: Prerequisites for *Kanban* Use

1. *Heijunka* production

When the production fluctuates too much every day, there cannot be a successful use of *kanban*. In the production planning extra large inventories will be created somewhere in the processes, including at the suppliers, if a thorough practice of *heijunka* in product mix and volume is not done.

2. Changeover time reduction

Just in time production has meaning only when one-piece flow or small lot production is practiced. Naturally, the number of changeovers increases. Therefore when companies attempt *kanban* without reduction of changeover times, production losses will increase.

3. Withdrawal by downstream processes

Kanban and products move approximately at the same time. Upstream processes make only what is withdrawn by downstream processes, when needed, in the quantities needed.

4. Defect rates are practically zero (defects are not be sent to downstream processes)

The basic principle is that the quality needs to be built in at the process. If there are too many repeated occurrences of defects, production planning becomes impossible and lots of losses will be created, such as work in process inventories and stoppage of production lines.

5. Machine breakdowns are practically non-existent

If there are too many incidents such as minor breakdowns of machines, production cannot cope with just in time.

6. Thorough Sort & Straighten (first 2 steps of 5S)

The *kanban* system starts from sorting and straightening. By sorting and throwing unnecessary things away, otherwise unnecessary space can be effectively used. Also, by straightening things, necessary things can be easily found and used effectively.

7. Thorough training (about people's awareness of waste)

Even by trying to eliminate waste, training needs to be provided as to what waste is. On such occasions, training is needed until people fully agree with the concepts and thorough awareness about waste needs to be implanted in their minds.

8. Thorough workplace-focus (gemba)

Problem solving needs to be attempted at the hot spot where the true culprit is found. It is necessary to establish a business culture with a workplace-focus, known as *genchi genbutsu* at Toyota.

9. Use *kanban* as a kaizen tool

If *kanban* either remains at certain locations or moves ahead of time it means we have some problems. Looking at the progress of *kanban* circulation, the problem points will expose themselves. Also, by looking after such problems visual management becomes possible.

Lesson 27:
Why Do We Practice 5S?

There are big differences in the operational fitness and strength of companies like Toyota which has been steadily and continuously doing self-study with strong eagerness to do *kaizen* and companies who have simply survived without making serious efforts. If the TPS is implemented in a company with weak operations it could be poison instead of strong medicine indeed.

Companies without such strong operational capabilities might be better off starting with the introduction of 5S (Sort, Straighten, Sweep, Standardize, and Self-discipline). However, it is necessary to know clearly what the purpose is before starting 5S. Just by cleaning, things can become cleaner for a time, but otherwise it will not last. We need to consider the number of man-hours required to sustain 5S.

Some companies clean their factories only when they receive customer visits. Their purpose is to give a good impression to the visiting customers. It would be good if they did the cleaning every day, but without having calculating what it means for man-hours, they don't implement 5S.

If 5S is implemented, the following points that had been unknown or that could not have been improved will disappear:

(1) Minor flaws

(2) Chronic defects

(3) Inefficiencies (such as the time looking for something)

What we need to be careful about at this point is that minute imperfection and chronic defects are very hard to detect. Defects in semiconductor plants cannot be decreased simply by sweeping the floor and removing the dust. In order to reduce the defects there it is necessary to have high-level clean rooms. The same is true far other types of problems.Further, in order to resolve the issues of minute imperfection a thorough and focused approach towards the goal is required.

Commonly accepted English "5S" terms and the Japanese equivalents:	
Sort	(整理 Seiri)
Straighten	(整頓 Seiton)
Sweep	(清掃 Seiso)
Standardize (Sanitize)	(清潔 Seiketsu)
Self-discipline	(躾 Shitsuke)

Figure 27: Definition and Purpose of 5S

Definition of 5S

① **Sort**
Technique to remove what is unnecessary
② **Straighten**
Technique to keep necessary things in available conditions, when ever needed
③ **Sweep**
Technique to keep necessary items (or places) away from trash or dirt
④ **Standardize**
Technique to sustain Sort, Straighten and Sweep practices
⑤ **Self-discipline**
Technique to make the habit of following the rule or what has been decided

What will improve with 5S?

The following points that had been unknown, or that could not have been improved will disappear:

① Minor flaws *
② Chronic defects
③ Inefficiency (such as the time looking for something)

As countermeasure

Implement thoroughly:

① For thorough implementation, in terms of manpower, it is necessary to have working hours calculated (manpower x time).

② Unless the purpose is narrowed down, thorough practice is impossible!

*Minor flaws:
 ① Usually, you didn't notice or even if you were aware, the imperfection was too minuscule so that you overlooked it.
 ② Each imperfection is unnoticeable, but when the number becomes large, productivity will be affected because of by the accumulation of small imperfections.

Lesson 28:
Sort is to Throw Away Unnecessary Things

In an injection molding factory, operators previously used different box wrenches to tighten or loosen bolts on the molds. When the production was urgently needed, the efficiency was low and the operators were frustrated without being able to do a good job.

The causes were studied and as a result it was found that imported machines were being used when the company was founded. However, because the machines were switched to domestic machines the component parts were using both metrics and inches. In terms of functionality, there was no problem, but subtle differences were there because foreign products were measured by inches, whereas domestic products by centimeters.

In general, "sorting" would mean that we separate things into different boxes, individually for the items in inches and the others in centimeters, before using them. However, the sorting work in this case example required that we remove all the imported bolts. This was helpful not only for the changeover time reduction, but also for the reduction of problems, such as damaging the mold.

As in this example, even if certain things are usable per se, if we clarify the purpose for each item these could become unnecessary things. In such cases, unless you get to the point of actually removing these things big results can hardly be reached. By throwing unnecessary things away, new usable space can also be created.

In terms of the disposal of items, this needs to be kept in writing: who was responsible, who was the decision maker, and the proper procedure needs to be followed, in accord with company policy.

Those companies which cannot practice sorting often have machines that probably cannot be used any longer but are kept with care. **Sorting means to throw things away**. The point of the technique in disposing of unnecessary things is that **even usable items per se need to be thrown away, if such items will never be used in the operations the company needs to execute.** Simply separating boxes for items into inch and centimeter would not have visible results.

Figure 28: Why is Sorting Necessary?

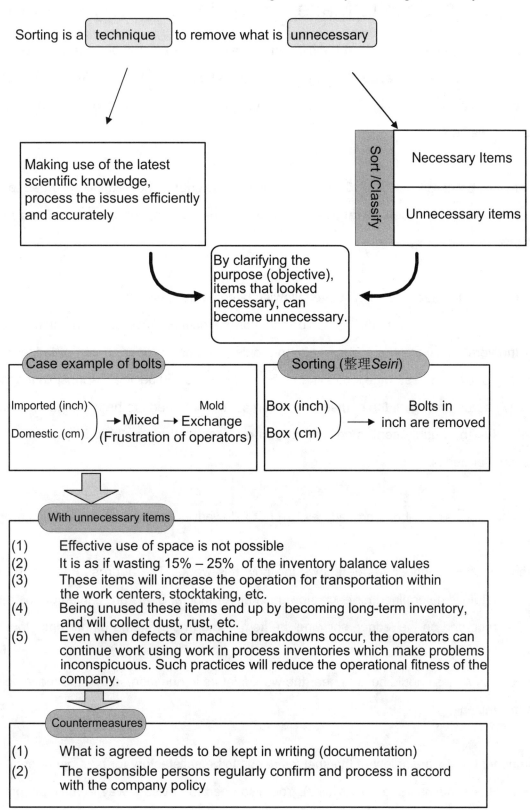

Lesson 29:
Straighten is to Make Things Immediately Available

An experienced operator went to get a mold which was not at the usual place and therefore he had to look in other places. He lost about 3 minutes just looking for the mold. At the time of mold changeover, necessary tools were missing in the toolbox. He then had to go to the tool room, which created another 2 minutes of waste.

If you are not bothered by about 5 minutes of waste looking for something, you cannot claim that you are practicing straighten. **Straighten is a technique to keep necessary things in available conditions, whenever needed.**

To do straightening, practices such as the following become necessary:
(1) Dispose of unnecessary items, (2) decide the storage location, (3) display the storage location, (4) identify the item (for example carving seals on molds), (5) keep proper control of ledgers to have no discrepancy between the storage location and the actual items, (6) decide responsible manager for the storage, (7) agree among all the parties concerned as to what needs to be continued for straightening (unless this is thoroughly confirmed, things can go back to the way how it was before) and also conduct regular audits.

In any case, unless the above procedures are not followed thoroughly, things simply can go back to the original status.

The diagram on the opposite page shows the how this works. The address (indicating the storage location in detail) is determined for each storage shelf for molds. Seals are carved on the molds. Then the data is recorded on the ledger, showing the item name (mold), item number and storage location number.

Frequently used ones should be kept near the work centers in question so that the movement distance can be minimized.

For component parts, the height of the shelves needs to be satisfactory for both operators (appropriate height) and the supervisors (easy visibility). You also need to devise a way to maintain proper space, so that no more than a reasonable amount of items can be kept there.

Figure 29: The Steps for Straighten

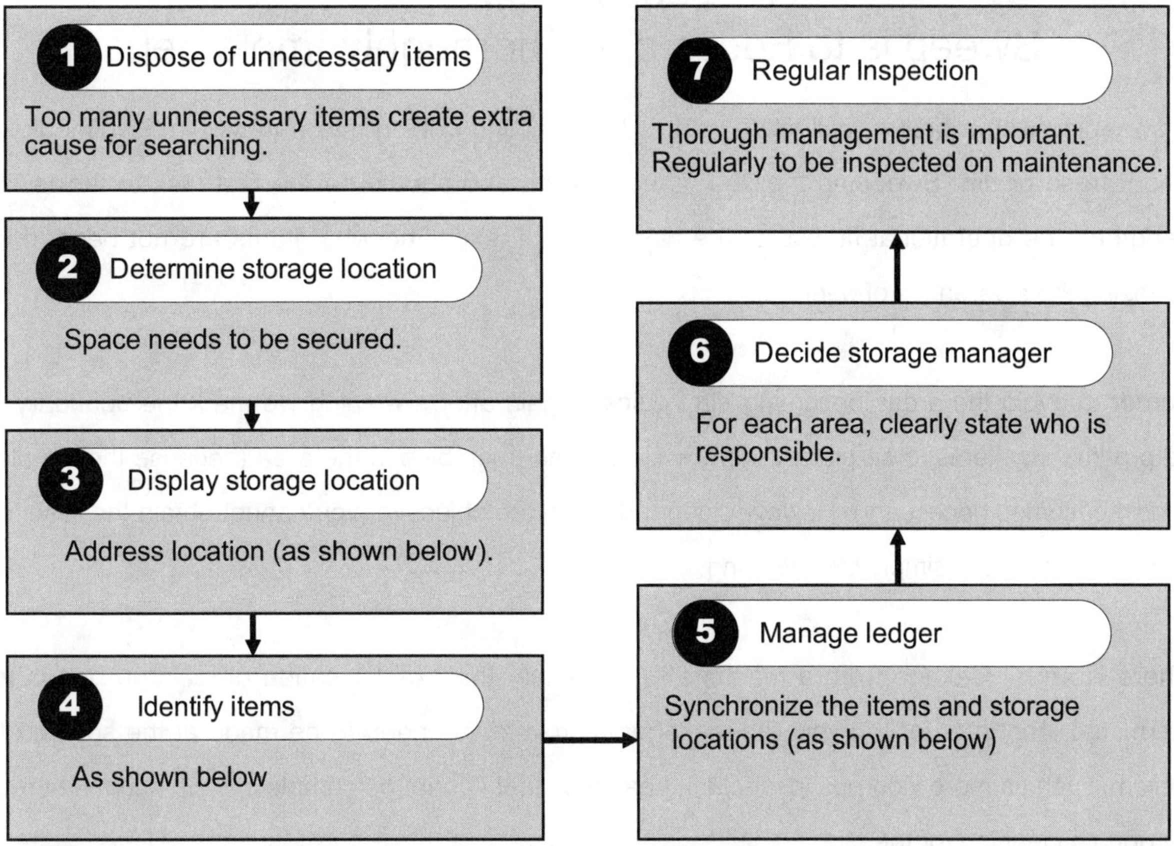

1 Dispose of unnecessary items
Too many unnecessary items create extra cause for searching.

2 Determine storage location
Space needs to be secured.

3 Display storage location
Address location (as shown below).

4 Identify items
As shown below

5 Manage ledger
Synchronize the items and storage locations (as shown below)

6 Decide storage manager
For each area, clearly state who is responsible.

7 Regular Inspection
Thorough management is important. Regularly to be inspected on maintenance.

Figure 29.1: An Example of Straighten with Mold Storage

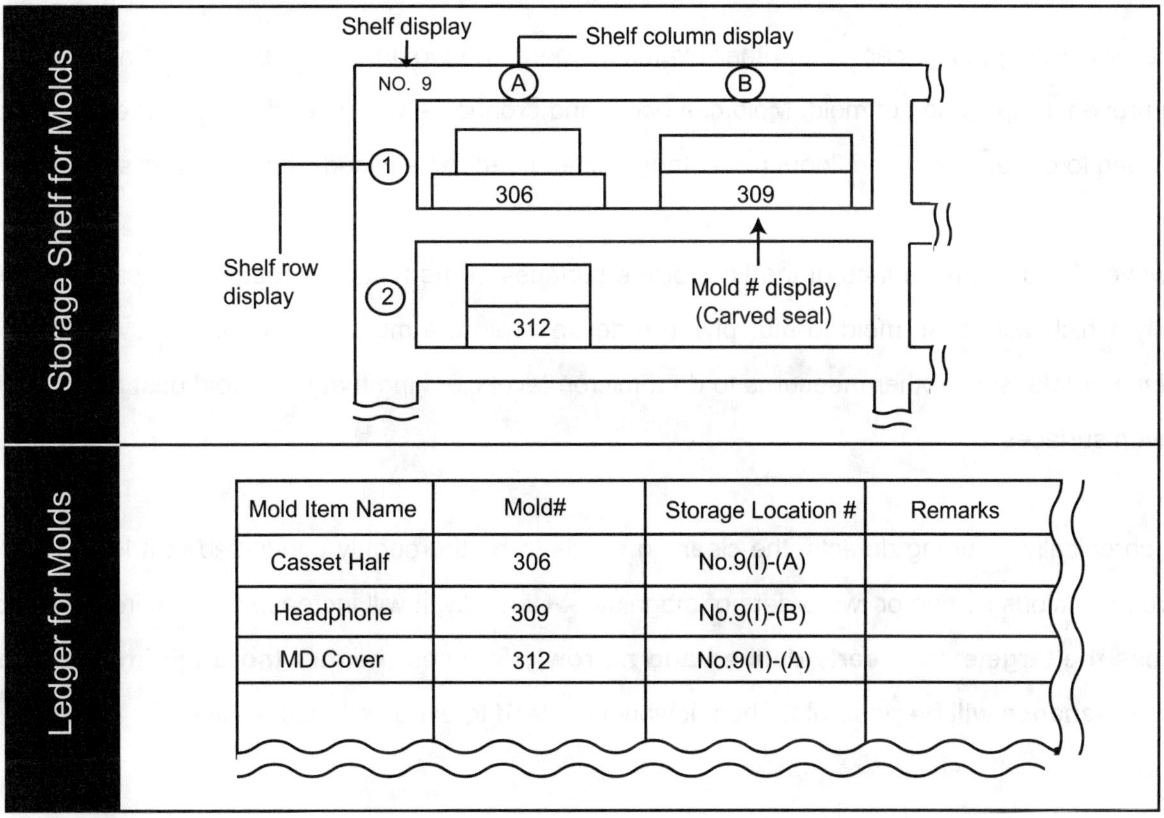

Storage Shelf for Molds

Shelf display — NO. 9
Shelf column display — (A) (B)
① 306 | 309
Shelf row display
② 312
Mold # display (Carved seal)

Ledger for Molds

Mold Item Name	Mold#	Storage Location #	Remarks
Casset Half	306	No.9(I)-(A)	
Headphone	309	No.9(I)-(B)	
MD Cover	312	No.9(II)-(A)	

61

Lesson 30:
Sweep is to Focus and Thoroughly Implement

By sweeping with a broom, collecting with a dustpan, and putting into a trash bin, it will look clean without trash or dirt. Sweeping the floor is easy to do and should be the first task to tackle. It will strengthen the operational fitness of the workplace by exposing what items are not needed while increasing the awareness of operators.

In order to avoid the areas becoming dirty quickly again after sweeping we mark the boundary lines and prohibit any items to be placed there while placing trash bins in the area (because the separation of trash becomes necessary). However, depending on the objective, you cannot obtain the results you might have expected simply by sweeping.

If there is an oil leakage from a machine for instance, the exact location of leakage needs to be determined, and in order to avoid the leakage, improvements need to be made at the source of the problem, such as more vigorous inspection. Leakages that cannot be completely avoided might require appropriate cleaning for the oil.

Food processing plants can be another example where you would also need a cleaning method that can prevent propagation of mold. Mold can occur and propagate wherever dusts or dirt exist. Dust and dirt need to be removed, and floors (and other places) need to be wiped with antiseptic solution.

However, if either the surface of the floor or the surfaces of the belts are rough, dust can accumulate easily, which will cause mold to multiply. In order to avoid the multiplication of mold, either wax the surface or take some other measures to do a micron-level cleaning that can avoid dust and dirt on the uneven surfaces.

For chronically recurring defects, the cleaning needs to be thoroughly conducted at a level that is far more meticulous by one or two orders of magnitude. Naturally, it will increase the required man-hours. **Unless the targets are clearly defined and narrowly focused, neither thorough implementation nor sustenance will be possible**. Then, it would not lead to expected results either.

Figure 30: The Steps for Sweeping

Steps	Description
(1) Sweep / Clean with broom and dustpan	① Clean the floor, desk, shelf, wall, ceiling ② Cleaning an area close to people will result in the improvement of their awareness at the workplace ③ By cleaning, problems will be exposed
(2) Set marked boundary lines	① Mark the boundary lines, and clearly indicate the athways and storage areas ② Prohibit that any items be left there, such as work-in-process inventories, moving shelves ③ Display danger signs, etc. ④ Place trash bins (dividing up rubbish into burnable and non-burnable) and fire extinguishers
(3) Clarify objectives	① In order to do the cleaning, specific working hours (manpower x time) are needed ② In order to directly connect to the results, the objectives need to be clarified ③ Narrow down objectives and thoroughly attempt to do the cleaning
(4) Identify the sources (of dusts, dirt etc.)	① Search through the circuits (oil pressure / hydraulic, lubricating oil, air pneumatic pressure, electric, water circuits) ② Identify the inputs (people, materials, component parts, pallets, ducts) ③ Investigate the source spot (punching, cutting, discharged glue)
(5) Micron-level *kaizen*	① Remove roughness of the floor surface — Dust enters / Floor surface / Cover with wax painting material ② Enlarge the image with microscope (identify micron-level dust, mould, germs) ③ Determine numerical values of temperature, vibration, smell, light, noise, etc.

Lesson 31:
Key Points to Start Implementing TPS with 5S

The foundation of 5S is actually **3S (sort, straighten, sweep)**. For instance, whatever was effective with 3S against the defective adhesion due to spilled oil or greasy dirt needs to be continued and sustained.

The general practice for the sustaining is that **we standardize what we agreed upon.** This action is called standardizing.

Also, **in order to be able to follow what was agreed,** self-discipline becomes necessary.

However, Toyota's operational fitness did not become so strong because of practicing 3S or 5S. The fact is that each process had the philosophy of receiving only "what is needed, when needed, in the quantities needed" so the objectives were narrowed down and what was agreed upon was thoroughly executed. As a result, the 3S improved.

For instance, if unnecessary items are left at the side of production lines these will be in the way of operations when operators transport or pick up something.

Then, people started to ensure that unnecessary items be never left in the way and agreed to certain things such as the addresses (storage locations for items in detail) so that necessary items can be procured at specified locations without any loss in time and motion.

The procurement of items can be based on the idea that no extra seconds of waste should be created. Thus, detailed procedures need to be firmed up with the agreement of all the parties concerned and that these need be executed accordingly. In order to be successful with 3S, this level of thoroughness is required, but the number of man-hours allowed within the company is limited.

We must remember that even if we call loudly for 5S or 3S, unless the focus is sharpened, thorough implementation cannot be achieved.

Figure 31: Doing 5S Does Not Guarantee Improvement

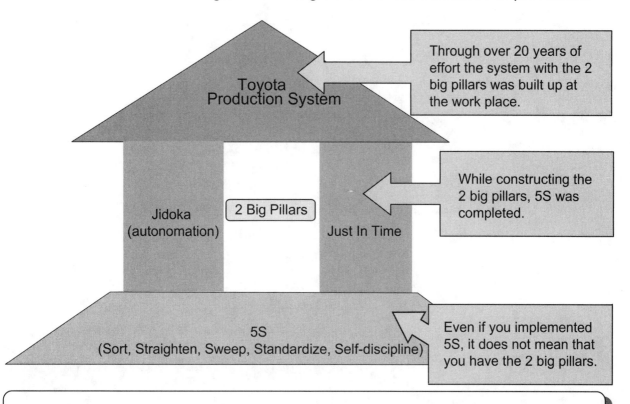

An increasing number of companies are starting over with 5S because they realize it is impossible to implement TPS without this operational fitness and culture.

Figure 31.1 Key Points for 5S Implementation

① 5S cannot be implemented aimlessly or randomly. Rather, it can then create waste.
② In order to succeed at 5S it is important to narrow down the purpose and to thoroughly execute.
③ For thorough execution, man-hours are needed. Unless targets are focused, it will not be connected to results and will create waste.

Example:
Move old documents to the right side, and dispose them when they do not fit into the shelf.

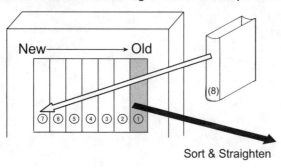

New document # 8 is to be posted on the left side of document # 7. Then, the document # 1 is to be disposed.

Methods of disposition)
① Throw away
② After discerning the contents store in warehouse, etc.

Section 3:

Jidoka,
Or automation with human intelligence (autonomation)

Lesson 32:
What is *Jidoka*?

As mentioned earlier, TPS consists of two pillars: just in time and *jidoka* (autonomation). This has already been mentioned. Just in time considers the flows between processes as something like a flowing river. It aims at maintaining flow without wastes, time gaps or stagnant areas.

Jidoka on the other hand, focuses **on stopping the flow.** At Toyota, decision making based on *human wisdom* is given to machines. This is done by building in a mechanism of making decisions as to whether things are good or bad and when problems occur sensors detect and stop machines or equipment.

This is the reason why it is called automation with a human touch or with human intelligence. It includes the process where the operators themselves push a button to stop the production line.

Stoppage of machines means that the line will stop. If the line stops, immediately *kaizen* activity should be started on the spot where the problem occurred. Because the improvement is made quickly within the process, it can prevent defects from being passed on to downstream processes or end-customers.

In the case of "automation without human intelligence," people for supervision would be required for machines with frequent trouble. However, automation with human intelligence does not need people for machine supervision because the machines stop when problems occur. This **enables flexible manpower** by continually optimizing the number of workers to meet the demand.

Also, because the problems are tackled on the spot without a time gap, true causes (culprits) can be more readily identified and lead to a solution. The *"in flagrante delicto"* (culprit caught in the act) is quite often easy to apprehend and the mystery can be solved without the need for high-level techniques.

Figure 32: Automation Without and With Human Intelligence

	Without	With
Stopping the machine	Production continues unless somebody switches off a machine.	Abnormality is detected by machine itself and the machine stops by its own decision.
Production condition	Defects are produced continuously.	Defects are not sent to downstream processes.
Solving problems	The search for causes is conducted later, and it is difficult to grasp the true causes.	Problem is tackled on the spot where it occurs and it is easy to grasp the true causes.
Eliminating waste	Lots of waste occurs, including at inspection of downstream processes. In many cases, you cannot reduce manpower.	Saving, adjusting and optimizing the use of manpower becomes possible.

Figure 32.1: Workplace-focused Rapid Problem Solving

I. Problem occurs
II. Communication from the place where the problem occurs
III Stopping machine/equipment in which the problem occurs
IV. Thorough practice of a workplace-focused philosophy (gemba)
1. *Genchi* (actual location): Go to the place where the problem occurs (supervisors, supporting staff, etc.).
2. *Genbutsu* (actual item): Thoroughly look at the location on the item where the problem occurred.
3. *Genjitsu* (actual fact): Observe well the conditions as to how the problem occurred and listen well to the operators.

Main Points

① Necessary to learn the technique of how to observe the actual location (genchi) and the main constituents.

② Discard preconceived ideas for problem solving cultivate the habit and persistence to repeat the why questions at least 5 times (5 Whys), until the problems are resolved.

Lesson 33:
The Relationship between *Jidoka* and Just In Time

Just in time smoothes the flow of the items without waste and removes time loss. *Jidoka*, on the other hand, focuses on stopping the flow of items. Thus, they might look quite contradictory.

Indeed, when the machine is stopped and the production line stops, the flow of items is disrupted. However, the reason to stop the machine is to swiftly solve problems at the point where these occur rather than leaving the problems until later. As a whole, this will make the flow smoother. By thoroughly practicing *kaizen*, we can establish a more efficient production system without waste.

Toyota took many years to make such improvements in eliminating wastes in the production processes. Over those years, these two concepts of just in time and *jidoka* have been formed in a unified way as the two main pillars.

Because of highly efficient machines and the mass production system, even after defects occur we would be continuously making defective products when we rely only on humans. Through *jidoka* it is possible to detect errors or defects immediately on the spot and avoid making large quantities of defects.

With *jidoka*, when an abnormality occurs the production line stops right away and *kaizen* takes place, as changes are made or measures are taken. However, many companies continue without stopping the production process while they take the data on defects or shoot a video on the condition of defect occurrence, postponing *kaizen* and correcting errors.

Although you more effectively guarantee quality products by stopping the production line as Toyota does and by starting *kaizen* to make immediate changes or by taking countermeasures, many companies do not do that because they have no such structure within their operation to respond expeditiously. Even if they stop the line, huge losses will incur for improvements and restarting the operation.

> "A production line which does not stop is either wonderfully good or quite bad."
> *Taiichi Ohno, founder of TPS*

Figure 33: Operational Improvements that Enable *Jidoka*

5 Whys

At Toyota instead of usual 5W1H (when, where, who, what, why, and how) the true causes are sought by repeatedly asking "why?"

Case example
(1) Q: "Why do we have metallic powder entered into the product?
A: "It entered within the drying process line."
(2) Q: "Why did it enter within the drying process line?"
   A: "Because the temperature in the conveyor line is high, the bearing can get easily damaged."
(3) Q: "Why do we have more of the metallic powder entering in Line C than other lines? We do have five similar production lines."
   A: "When we measured the temperature, it was 30 degrees centigrade higher in this line."
(4) Q: "Why is the temperature of Line C high?"
   A: "The burner's angle, position, and inspection frequency were different from other lines."
(5) Q: "Why do the other lines have the metallic powders entered, even with less frequency?"
   A: "The bearing deteriorates, which is inevitable. Before it becomes damaged the bearing needs to be replaced. We should have regular inspection using sound scope hearing and when there is an abnormal sound which is a sign of damage, we need to replace the bearing."

In this case, if asking why four times only, then we still have metallic powder entering products. By repeating the why 5 times we reduce the frequency of bearing replacement and reduce the metallic powder in products as close as possible to zero. To repeat asking why five times means that the true causes need to be thoroughly sought.

Basic Ideas about *Jidoka*

① Constantly think of ways to separate manpower from machines and recognize that people are more important
② Maintenance of machines and equipment
   Regular checkups (oil pressure, air pressure, blade replacement, oil replenishment, other quality related issues, etc.)

Lesson 34:
Why Do Companies Not Succeed at *Jidoka*?

Jidoka focuses on stopping the production line and is a mechanism in which the line stops when abnormalities occur.

The line stops because of abnormalities, and therefore cannot work if there are too many of the following abnormalities:

(1) Frequent minor breakdowns of machines or equipment

(2) Frequent defects whose causes are not identified, and without the capability to build quality in at the processes

(3) No staff have been trained to identify the causes and make necessary improvements

(4) No system exists that can detect abnormalities on a real-time basis and take swift countermeasures

(5) Without standard work there is no mechanism to stop the operation when there is an abnormality as a deviation from standard

(6) When the line automatically stops there is no device to communicate the situation, such as *andon* or an alarm system.

(7) Setup time reduction has not been promoted

Unless the issues such as the above have been improved at a high level, the production line frequently stops and no swift countermeasure can be taken. Then the stoppage time becomes long. This will lead to the inability to secure a required production quantity and the fear that delays would occur in deliveries to customers.

When the line stops, it means that the quality is often unstable at the next start of the operation and that until the quality stabilizes you would be making defects. The more frequent the stoppage the bigger loss in the yield.

> The *andon* is a lamp that indicates the status of an operation that is often used in combination with pull cords to enable workers to signal when there is a problem.

Figure 34: A Production Line Capable of *Jidoka*

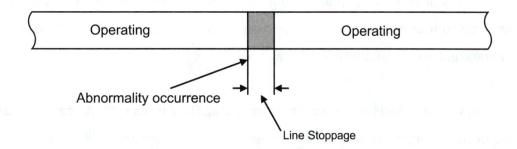

Abnormality occurrence
Line Stoppage

① Stoppage is not frequent
② Capable of swiftly recovering the operation (including fine tuning adjustment)

Advantage
Within the time lag, the cause is sought and it is likely that true causes are found and solved.

A Production Line Barely Capable of *Jidoka*

1. It takes too much time to recover the operation
2. Not capable of identifying causes

Operating | Operating | Operating | Operating | Operating

■ Unstable quality after the re-start of the operation be inspected ⎯ (a) Items need to
▨ Line stoppage   ⎯ (b) Yield erosion

Problems
① Frequent stoppages
② Mostly quick-fixes are made, not becoming capable of recovering the operation

Lesson 35:
Quality is Built in at the Process

In any company, the goal is to have **quality built in at the process**. In practice however, many companies avoid doing that by resorting to inspections. Inspections, even done efficiently, do not create added values and simply create enormous waste.

Toyota's idea is not to find defects but to **make no defects**. They thoroughly practice the idea and when problems occur, the machines detect abnormalities and the production lines are stopped by machines or by operators.

In order to stop the production lines, it is desirable that the **countermeasures against defects are thoroughly practiced** and the defects are practically zero. For that purpose, the operation needs to established a workplace-focused culture and of solving problems by getting the facts at the place where the problems occur.

When a problem (defect) occurs, the true causes are sought on the spot while it is still occurring so that the problem will not recur, just as culprits can be easily apprehended without the need for advanced investigation techniques if caught red-handed. When defects occur, the ocecurrence is the chance for problem solving and can lead to continuous improvements.

With **workplace-focused problem solving approach**, the gemba (actual place of work) is still unchanged and the point is how swiftly you can act when the problem occurs. Even if the people come right away, it can be that the true culprit escapes (the cause of the defect disappears) and the causes cannot be identified.

In such a case, if the operators who are always in contact with defects have accumulated the habit of asking themselves why five times (the 5 whys), problems can be solved one after the other. In this way defects can become practically zero. The diagrams on the opposite page illustrate the writer's problem solving method based on workplace-focused approach.

Figure 35: Workplace-focused (gemba) Problem Solving

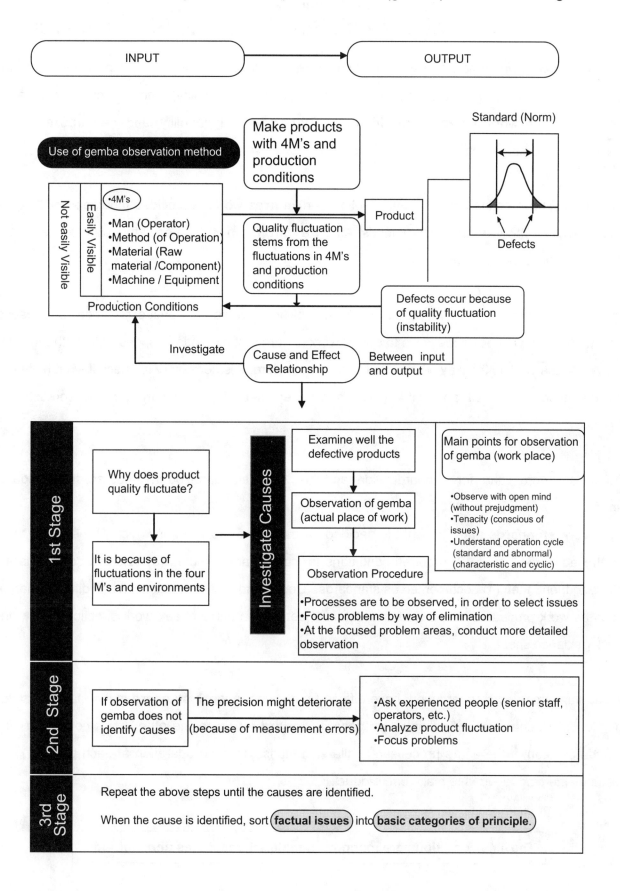

Lesson 36:
Differences between Work Standards and Standard Work

The reason why we standardize work is because we aim to make continuous improvements on the actual operations where we can find problems (fluctuations or variability). The problem is that the **4M conditions and work environments fluctuate.** The problems (variability) and the purpose of work standard are illustrated on the opposite page.

Standardization means that we want to follow certain rules we have decided together. For that, we select what we think are the best methods. After we start, we might see something is inconvenient or defective. Then, we can revise it.

In practice, in order to avoid the problems and variability 1) we sort our past experiences and results, based on which we decide the ways to do the work (in terms of methods, responsibility, authority, etc), (2) We do the work in the way we decided and agreed. Here, it is necessary to check if what has been decided is being followed. and,3) if anything is inconvenient or defective in the way the work is done, we revise it.

Above is the procedure for standardization, and the written standardized methods are the standards sheets.
The work standards sheet is what is decided on as the way to work, in which 1) the items to be controlled are selected, (2) required conditions for control are selected (3) selected conditions are stabilized, and then (4) documented as standards. Depending on the company, it has different names such as work procedure sheet, work instruction sheet, work outline sheet, work direction sheet, and work guidance sheet.

Work standards and standard work have different meanings. A **work standard** is the standardization of the work done in each individual process. **Standard work** on the other hand constitutes the standardization of the entire processes (while encompassing each operation at each process) and includes the time required to make one product.

> The 4Ms conditions are manpower, material, machines and methods.

Figure 36: The Purpose of Preparing Work Standards

Problems (Variability)	Purpose of Work Standard
① If the 4M and environments vary, the result (product quality) varies.	Decide the work method.
② New or inexperienced workers tend to vary in their work (variance), and need more time (time variance).	Enable new or inexperienced workers to do the same work within a short period of time. Training to be made easy and efficient.
③ Depending on the person, there are different ways of instruction.	Instructions and supervision of work need to be made easier. so that the work can be executed efficiently.
④ The current work is not clearly grasped, and therefore it is not efficient.	Current work needs to be reviewed. and clearly documented, so that the work becomes efficient.
⑤ Assigned share of the work is not clear and each operator is different (varies).	Current work needs to be clearly shared to make sure that each operator ompletes the assigned work.
⑥ Unclear responsibility and authority of each section.	Clarify each section's responsibility and authority.
⑦ With high variability / low volume, waste is created in inventory, setup, etc.	With standardization, increase compatibility and general purpose use.
⑧ Skills are within individual's experiences, and things need to start from the beginning, if that person is not there	Sort the skills and data that are with individuals. so that anyone can use them .

Figure 36.1: Work Standard and Standard Work

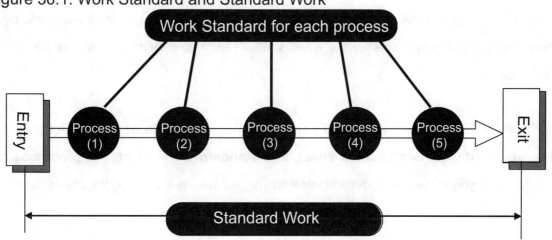

Lesson 37:
Key Points for Standard Work

With *jidoka*, the machine is stopped when abnormality occurs, but we need a standard or reference point for the abnormality. This standard is the standard work sheet.

In case of machines or equipment, the mechanism is built in the machine to make decisions on what is good or bad in order to stop the production line. So there is no issue here. However, for operators we need standard work as the reference point for the decisions to be made.

(1) Takt time

It is the time to make one piece of product and represents the operation speed. The takt time is based on the quantity needed by customer.

> **Takt time**
> = Available time for production per day / Required quantity to be produced per day

(2) Work sequence

This is a detailed description of operations executed by each operator. For each operation, time is also shown so that whoever does the operation the same sequence and time will be followed.

(3) Standard work in process

This is the minimum quantity of necessary work in process inventory for each operation. Those loaded on machines and items curing or taking time to dry are also included.

Standard work is based on the above three elements, and further improvements are made by adding operation outline sheet (key points for the work, special remarks, experienced worker's know-how as standard sheet), etc.

In order to establish standard work, you need to prepare the **table of production capacity by process, standard work combination sheet, and standard work sheet**. The production capacity document is to let you know the production capacity by process, as shown on the opposite page.

Figure 37: Takt Time

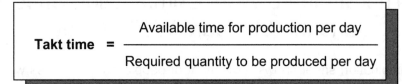

N.B.
(1) Time for production per day is the regular time and does not include over time.
(2) Required quantity to be produced per day is the required quantity for the month divided by the number of working days of the month.

(Example)

$$\text{Takt time} = \frac{400 \text{ minutes/day}}{510 \text{ pieces/day}} = 47 \text{ seconds/piece}$$

Figure 37.1: Table of Production Capacity by Process

		Table of Production Capacity by Process									
Table of Production Capacity by Process		Part #		Part Type			Approv By		Date		
		Part Name		# of Pieces			Confm By		Prep'd By		
Step #	Process Description	Machine#	Base Time				Tool Change		Total Capacity	Remarks	
			Machine		Auto time		Machine Cy		Pcs per Change	Time to Chnage	
			Min	Sec	Min	Sec	Min	Sec			
1	Pickup Raw Mat			3				3			psc
2	A			5		23		28	100	2'50"	1,000
3	B			6		23		29	80	3'	1,000
4	C			8		30		38	15	40"	800
5	D			4		37		41	10	30"	1,000
6	E			4		32		36	10	20"	750
7	Place Processed Material			3				3			
	TOTAL			33							

Remarks: (1) This becomes the standard for the preparation of Standard Work Combination Sheet
(2) Total capacity is the capacity within the regular hours of the day

Lesson 38:
The Standard Work Combination Sheet and the Standard Work Sheet

A standard work combination sheet shows movement of people and machines within the takt time at a glance. As on the opposite page, even if we have seven processes the work balance can be seen clearly. Thus, problematic processes can be easily identified, leading to continuous improvement.

Accurate times are checked, such as manual work of loading parts on machines, walking from machine to machine, and automatic material transfer by machine. Operations continue in cycles, so one work needs to be completed before the next cycle starts. The point for improvement is whether we have extra surplus time at that point (when one cycle has ended and the next should start).

A standard work sheet is what shows the operation scope for each operator. This is prepared so that anyone can see the contents of the operation.

With repeated operations, one cycle time becomes the takt time. The symbol (●) in the Standard WIP field shows the number of standard work in process pieces. Machines using blades are shown with the symbol (+) meaning the safety precaution is needed at that process.

Thus, as for the operations for which operators' motions are key factors (instead of machines), standard work is prepared for the whole processes. When the whole processes cannot implement the standard work concept, then as much as possible, the standard work combination sheet concept can be implemented in order to increase the level of waste elimination.

When standard work is prepared, training is necessary in the way the operators understand and agree. Also it is to be noted that not only when the line stops but also when the line is moving smoothly, each operation keeps being continuously improved.

> **Symbols used on the Standard Work Sheet:**
> ◇ = Quality check
> ✚ = Attention to safety
> ○ = SWIP inventory

Figure 38: The Standard Work Combination Sheet

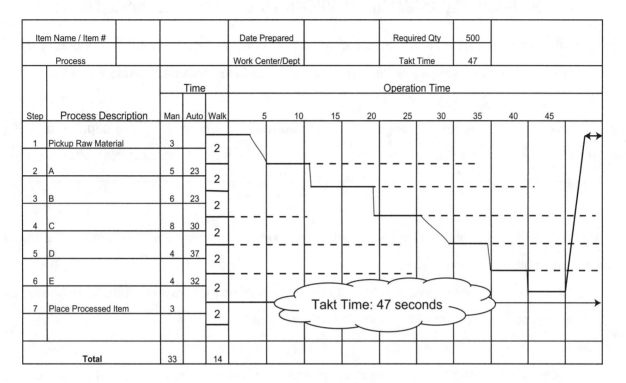

Figure 38.1: The Standard Work Sheet

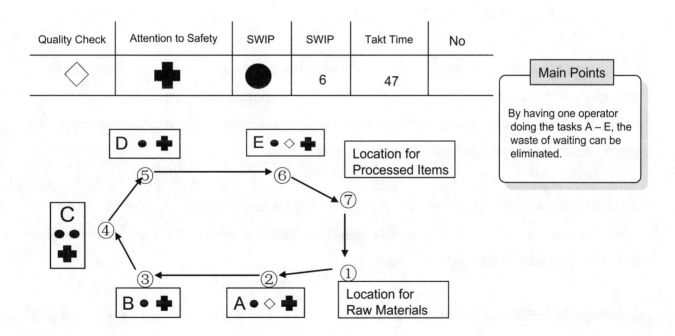

Lesson 39:
Reasons Why Standard Work is Difficult to Establish

In order to run your operation based on standard work it is necessary to make your working place capable of repeatable operations for all the processes with the focus on human motion.

However, even if some processes can be standardized not many companies are capable of using standard work for all of their processes because various issues as in the examples shown on the opposite page.

The Toyota Production System (TPS) is based on teamwork. The teamwork we are talking about consists of quite highly skilled operators.
In such sports as baseball or soccer, a team with one or two superstar players tends to be beaten by other teams with good teamwork without high-level superstars.

There are companies seeking advice, explaining that even with serious attempt at implementing the standard work with the three elements (takt time, work sequence and standard work in process inventory) their efforts did not result in profits. The reason for this was that their efforts were applied to only one part of their operation.

By examining the details, we would find that although the processes are structured indeed by the three elements, the processes are not linked. Often, busywork (apparent operation) occurs in other processes, or some processes are much too loaded and as a result the morale of the working place is lowered and the teamwork spirit is not there any longer.

In that way you can never achieve productivity improvement or cost reduction. For that type of situation we must thoroughly examine the current problems and start to prepare the standard work which matches the level of the company's operation.

The concepts of TPS are excellent and can be good examples as a target direction for your operation. However, if you implement these without being aware of the level of your own operation, it is a fact that you may end up with more problems.

Figure 39: Remove Variability before Implementing Standard Work

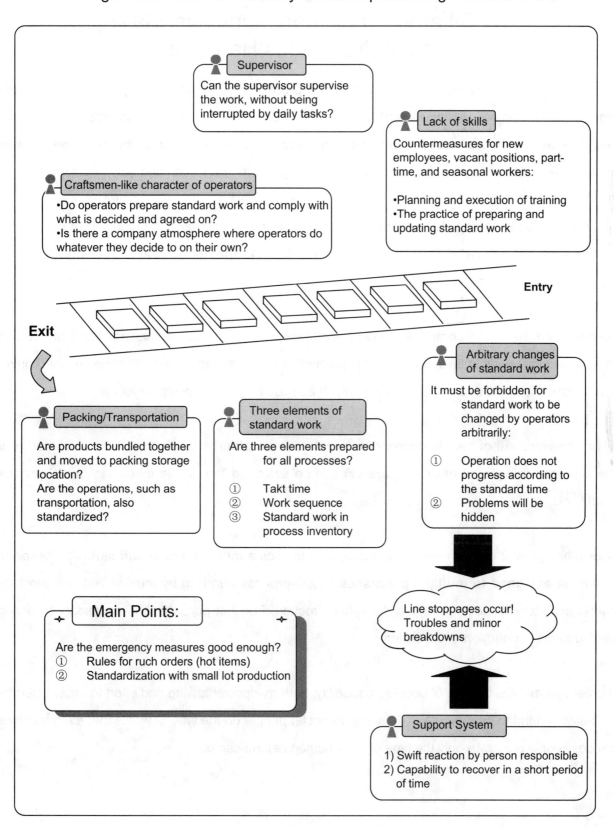

Lesson 40:
The Difference between Manpower Saving and Multi-process Handling

With *jidoka* if machines stop automatically when problems occur, those people who are solely in charge of watching the machines work become redundant. We can then reduce manpower. At that time, if we reduce the workload by 0.5 persons, it will not lead to manpower saving. **Manpower saving** is to reduce manpower by the unit of one person through continuous improvement.

Also even if we reduce the manpower by the unit of one person, if that one person does not do other work, we cannot say we did continuous improvement. **Manpower saving** means that the saved manpower is utilized somewhere else.

Even with *jidoka*, some companies have surveillance people to see if the machines are running properly. *Jidoka* needs to be improved to the point where such supervision of machines is not required. Toyota makes effort to avoid such waste as much as possible, which results in *jidoka*.

When the workload of people decreases, people can do other work. An operator who used to do only one task **(single functional operator)** can be assigned to multiple machines **(multi-process handling)**.

Also, if the operator can do many other tasks, he becomes **a multi-skilled (multi-skilled) worker** and he can be **assigned to multiple processes**. Multi-process handling by multi-skilled operators can lead to inventory reduction and is a desirable direction. For that purpose, it is necessary to provide training to make operators multi-skilled.

In order to maximize the use of people's capability, both manpower saving and effort to make operators multi-skilled need to be promoted. Thus, we do not let people do the type of works that even machines can do. People are given only the type of work humans alone can do.

Figure 40: Manpower Saving and Multi-process Handling

Manpower saving method

Manpower saving is to reduce by the unit of one person through continuous improvement. The main point is to separate the roles of operators from those of machines. A similar concept is labor saving, which is simply a replacement of manual operations with machines.

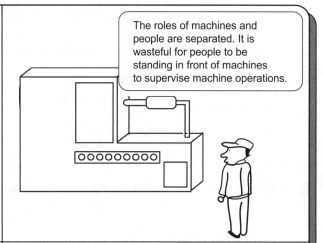

The roles of machines and people are separated. It is wasteful for people to be standing in front of machines to supervise machine operations.

Multi-process handling

One operator is assigned to multiple processes within the same takt time. It is also called vertical handling, originally used for logistics' vertical transport of materials between different floors of a building.

(Characteristics)
- Make one piece of product at a time.
- Easy to level the load (*heijunka*)
- Does not increase inventory
- Strive to have more multi-skilled operators.

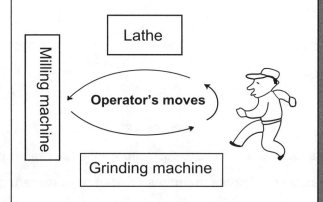

Multi machine handling (operators assigned to many machines)

One operator is assigned to multiple machines that have similarities. It is also called horizontal handling, originally used for transport of materials to the next location or process.

Make operators multi-skilled

One operator is assigned to multiple machines that are of different types.

-> For this, training is needed and charts are used to keep track of operators' training records.

Capable of teaching it

Capable of following instruction or plan

Capable of responding to abnormalities

Capable of doing it on their own

Lesson 41:
Flexible Manpower

As in a flowing river, when obstacles to the flow get removed one by one, smooth flow or "flow production" can be achieved. Then, the most effective distribution of man-hours (=manpower x time) is to be achieved for the line.

The man-hour reduction process designs a production line that can maintain the same level of productivity, in response to variable production demand, by adjusting the manpower.

Customer demands fluctuate. When production quantity decreases, if you operate with the same number of people (the system of a fixed number of operators, without adjusting the man-hours), then the productivity decreases. In such a situation, it becomes necessary to operate with less number of operators.

When the operation processes cannot be separated from the main production line and the man-hours cannot be adjusted, such situation is referred to as "**remote islands**" which hinders the **flexible manpower. Flexible manpower lines focus on reducing man-hours (= manpower x time) and is a different concept from manpower saving.**

The manpower saving eliminates wastes by focusing on manpower reduction techniques. This enables the operation originally performed by five operators to be done with three only. On the other hand, man-hour reduction is intended to enable flexible production even for the reduced production demand quantity: the line can operate with either three or four operators. Through flexible production, waste can be removed and further reduction of manpower becomes possible.

In order to focus on man-hour reduction, certain prerequisite conditions as shown on the opposite page need to exist. If such prerequisites are not satisfied it can lead to over-intensified labor and fatigue. Then the fatigue can increase the risk of endangering the safety and good quality products cannot be made either. Thus, on a daily basis it is necessary that the operations satisfy such prerequisites.

> **A process is known as a remote island when it is physically separated by distance or isolated from its upstream and downstream processes. This makes it difficult to adjust to variations in customer demand by combining and rebalancing work tasks for flexible manpower.**

Figure 41: Prerequisites for Flexible Manpower

Prerequisite	Description
Reexamine the existing production system from the foundation, and adopt the flexible manpower method.	① Stop lot production and start one piece flow ② Do not limit the work operators can do
Flexibility to adapt to the production line changes.	① Rather than big robotic machine implementation, make effective use of manpower who have greater flexibility ② Do not fix the machines, equipment, tools, to be able to easily adapt the processes to the production line variation
Elimination of waste, such as moving, by examining the layout	① Enable one-piece flow, by placing machines/equipment in the same sequence as the production processes ② Eliminate separate "remote islands" and gather operators at one location, so that team work is possible ③ Layout by process, U-shape layout of machines
Promote development of multi-skilled operators	① Enhance work rotation and train operators to be multi-skilled ② Assign operators to multiple tasks ③ Promote standardization of the operation, and avoid unbalances among operators' operations
Calculate the takt time and try to assign required number of people, by allocating the number of working hours in the most effective way	Utilize the takt time, by adjusting the number of operators in response to the variable demand

"Remote Island" Work Sites and Flexible Manpower

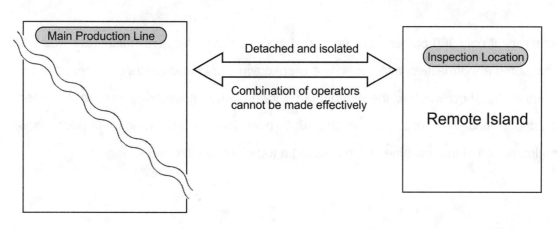

Lesson 42:
The Difference between Rate of Operation and Operational Availability

Usually, the rate of operation is "the percentage of the production result compared to the machine capability, when the machine is fully utilized." If a machine that can make 200 pieces per hour makes 60 pieces per hour then the utilization rate is 30%.

Although the machine has the capability, if there are many changeovers or short-term breakdowns and defects are made this is a waste to be continuously reduced. When improvement can move in accord with its capability, the utilization rate goes up more and more.

Thus, if the machines can be fully utilized, in full operation during the working hour the production quantity will increase. The only issue is that it can have the "waste of overproduction."

At Toyota, when operating machines, the operational **availability rate is used to show whether the machines indeed operated as required by the customers (or downstream processes) demand quantity, to the extent of the capability**. For instance, if the operation time is 8 hours a day, a machine that can make 200 pieces an hour would produce 1,600 pieces by operating to the fullest capability.

However, if the customer demand quantity was 600 pieces a day the extra 1,000 pieces were the waste of overproduction which leads to the waste in inventory. If we operate the machine for three hours only, the availability rate becomes 100%. It will not exceed 100%, which is the maximum.

On the contrary, if only 300 pieces were produced in three hours, the availability rate was 50% which means that some abnormalities, minor machine breakdowns, or changeovers occurred. The availability rate is an index to show whether the machines operated, when it was required, to the extent of the machines capabilities. We aim to achieve the 100% availability rate for which it is necessary to have proper maintenance of the machine, while keeping a log of the breakdowns.

Figure 42: Rate of Operation and Operational Availability

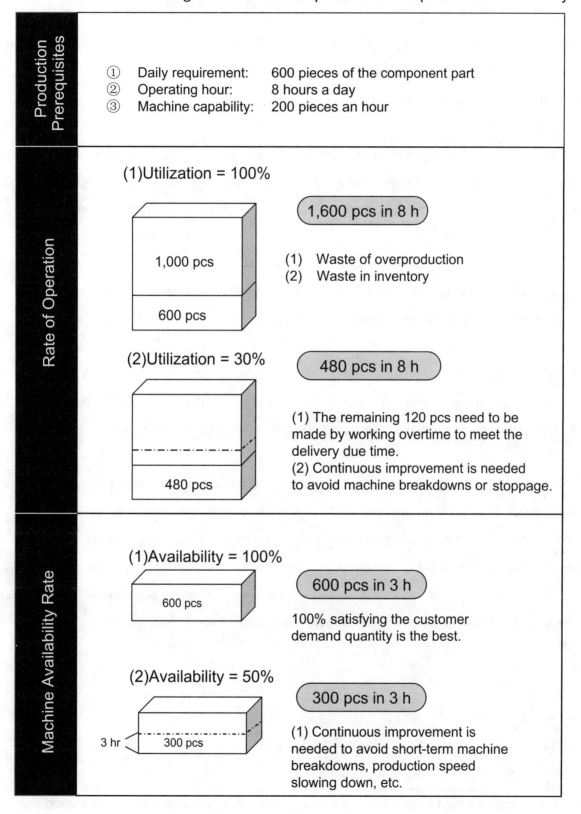

Lesson 43:
Key Points for Setup Time Reduction

Heijunka (production smoothing) is a method of production where only the quantity sold to the customers is made. As a result it becomes a high variety low volume production, and the number of changeovers (setups) increases for the item changes.

During the changeover, the production stops and even with your best effort in the changeover it does not create any added values. Because of small lot production the number of changeovers tends to increase, which can lower productivity. Nonetheless, it is possible to shorten the changeover time.

If you can shorten the changeover time you can easily cope with the production wish for reduced lot sizes and achieve the target of *heijunka*. Further, even if *jidoka* stops the production line, within a short time you can correct the problems, such as mold exchanges.

The changeover itself does not create value. It is best to make the changeover time as close to zero as possible. Because of advanced techniques, setup operations which used to take two to three hours became a single-minute changeover (less than 10 minutes) or a one-touch changeover.

The main point of the setup operation is to separate internal setup (the setup operation that cannot be done without stopping the machine) and external setup (the operation that can be done without stopping the machine). The actual steps are 1) first separate the setup operations into internal and external, and 2) change as many internal setup operations as possible into external. The rest is to do continuous improvements as shown in the diagram on the opposite page. Because of the advanced setup techniques, if following these steps the setup time can be shortened.

> **Internal set up tasks must be performed while the machine is stopped.**
> **External set up tasks can be performed while the machine is running.**

Figure 43: How to Approach Setup Time Reduction

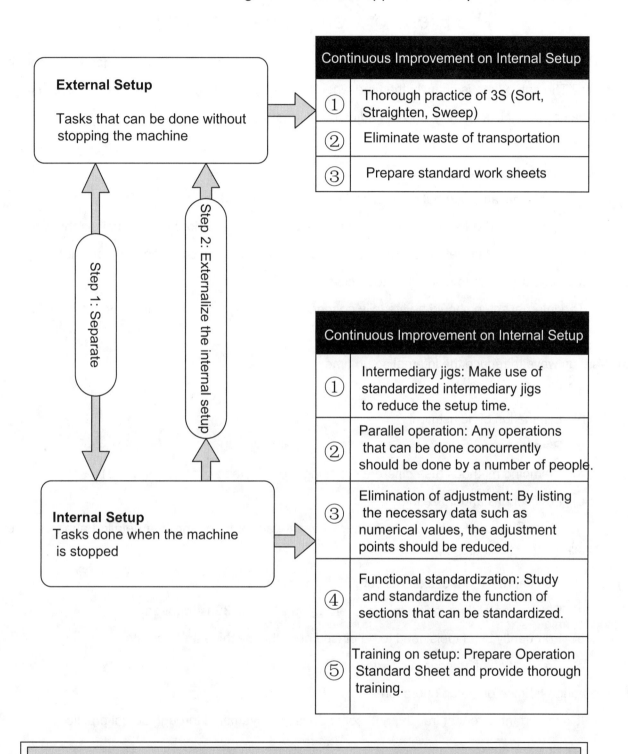

Lesson 44:
The Steps of Setup Time Reduction

Setup time is "the time between the last good piece off the current run and the first good piece off the next run." The setup time is actually the time when the machine is stopped, or not in production mode adding value even if it is moving, such as adjustment time.

The step toward setup time reduction are:

1) Separate internal and external setup activities

    Setups will be divided into internal setup (operations that cannot be done unless the machine stops) and external setup (operations that can be done without stopping the machine).

2) Convert internal setup to external setup activity

    The internal setup should be steadily changed into external setup as much as possible through continuous improvements.

3) Various ways to reduce the internal setup time

    a) Intermediary jigs

    b) Elimination of adjustment

    c) Parallel operation

    d) Clamping methods (use different clamping tools, instead of attaching bolts)

    e) Functional standardization

    f) IE (industrial engineering) techniques (waste can be found through detailed analysis, such as motion analyses)

4) Continuous improvement of external setup

    a) Elimination of time to find things (thorough practice of 3S is required)

    b) Frequently used tools need to be placed near the machine and operator

    c) Preparation of standard work document

5) Thorough training on setup operation

    It is important to repeat training and continuous improvements through actual practices.

Figure 44: Separate Internal and External Setup

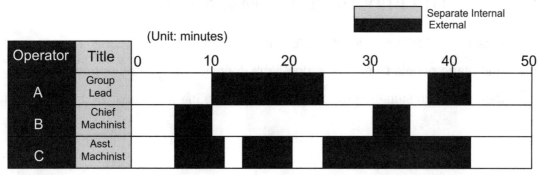

Figure 44.1: Eliminate Adjustments

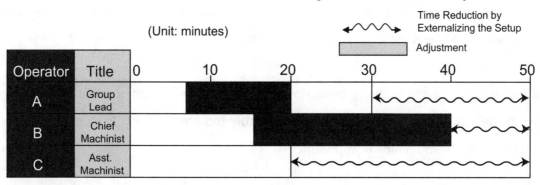

Figure 44.2 Conduct Set Up Activities in Parallel

The operators A and C help operator B at the bottleneck process, as soon as they are done with their own operations.

Various improvements
 (1) Intermediary jigs, (2) Functional standardization,
 (3) Industrial EngineeringIE (industrial engineering) analysis

Lesson 45:
The Importance of Maintenance

Within the *jidoka* practice of TPS when incidents like machine breakdowns occur, the machines detect the abnormality and alarm people using the andon lights (originally a type of lantern). People on the production floor then decide to stop the production line. Because the line stops, we need to rectify the problems.

If such problems are assigned to outside companies, we would spend extra time and money. Also, it is an issue of how quickly we can respond to the situation. Toyota provides training within the company so that such problems can be fixed. In that way, such operational structure has been established to recover the previous conditions within a short period of time.

However, if the machines have frequent breakdowns, production is not possible. In order to enable *jidoka* we need the machines and equipment which hardly require stopping the operation. For that reason as well machine maintenance is important.

Maintenance is an activity to maintain machine capability and includes inspection, supplying oil, repairing (and /or care) and adjustment. The maintenance type can be autonomous, breakdown, preventive, improvement, or protective.

Maintenance in general is done by machine maintenance staff. However, there are certain symptoms before the breakdowns such as abnormal noise or vibration. The operators who are in contact with the machines on a daily basis should receive trainings so that they can respond to such symptoms based on the idea of "**protecting our machines by ourselves**." The operators can also do their autonomous maintenance, such as cleaning and oiling their machines on a daily basis.

The importance of maintenance is more and more recognized by many companies to the extent that they implement Total Productive Maintenance (TPM) which are preventive maintenance activities performed by all the parties concerned.

Figure 45: Types of Maintenance

Breakdown Maintenance	• Only after breakdowns occur, the repair starts • If it is only after breakdowns occur, the time for repair takes longer and the productivity during that time lowers • Around the time of breakdowns, the product quality tends to be lower and causes delays in the delivery to customers
Preventive Maintenance	• Find abnormalities early and do the repair, before breakdowns occur • Have daily checkup and regular servicing (Good for autonomous maintenance and continuous improvements)
Improvement Maintenance	• When breakdowns occur, make improvements so that breakdowns will not recur, and also make continuous improvements for easier maintenance or repairs • Need to have an operational structure so that true information from operators will be collected, when the breakdowns occur
Protective Maintenance	• When planning new equipment/machines, prepare to have machines that will have fewer breakdowns, with high durability and easy maintenance • When breakdowns occur, the operators' real information will be given to those who designed/installed the machine
TPM or Total Productive Maintenance (preventive maintenance activities performed by all the parties concerned)	• Continuously improve the character and nature of people and machines • Operators themselves do maintenance • Machine maintenance staff used to do repairs when machines break down, but operators should act based on the idea of "protecting their machines by themselves" • Thus, operators need to acquire the skills for machine maintenance • Autonomous maintenance is included and the operators execute their daily checkup, 5S and other small continuous improvements, with the awareness of taking care of their machines • Preventive maintenance is extended to operators, and can be considered continuous improvements for people's motivation

Lesson 46:
Safety Takes Precedence over Everything Else

The prerequisite for the *jidoka* is safety. The basic philosophy is that "safety takes precedence over everything else."

Some people think that emphasizing safety is an obstructing factor to the improvement of productivity or continuous improvement of quality. This is a serious misunderstanding. At Toyota the belief is that "As we value our people, to that extent favorable results will follow." Productivity improvement without consideration of safety cannot exist.

For instance, if machines are running while someone is doing the cleaning, there is danger. Not only the cleaning tools but the edge of the operator's clothes can be caught in the machine. When the machine was running an operator was wiping the surface of the machine with a cloth and he dropped the cloth. He panicked and by trying to pick up the cloth, the edge of his clothes was caught in the machine. Usually, this would not happen but in haste people may move their hands unconsciously and cause an accident.

In such cases certain rules need to be prepared such as "no sweeping or cleaning when the machine is running," or "no hands into the machine during the machine operation," and thorough training should be provided. By recognizing the causes of accidents and by making a habit of following rules determined most incidents can be avoided.

With automatic machines there is a danger when an abnormality occurs that people take actions in haste and get caught by the machines. However, with *jidoka* (not automation), the machines stop because of the mechanism to automatically stop when an abnormality is detected.

Because supervision is entrusted to the machines themselves, no operators for machine supervision are needed and the danger is decreased. Therefore, the automatic stopping mechanism with *jidoka* not only results in quick actions for solving problems, but also is very useful for safety purposes.

Figure 46: Key Points for Safety

Item	Description
(1) Even a tiny accident should not be considered lightly and consciousness is needed to remove the causes	* Heinrich Principle of 1:29:300 Behind one major accident there are twenty - nine small accidents, such as scratches, and three hundred experiences in which people are not injured but have near-misses. → When the cause of an accident is found, even a tiny factor should not be left unresolved and countermeasures need to be considered.
(2) Make a habit of continuously improving before incidents occur	Including a tiny incident, you notice you need to start making improvements ahead of time without thinking: "This is not a big issue."
(3) Make a mechanism to remove careless mistakes	Avoid accidents even if people are sometimes careless or absent - minded. → Some safety mechanism needs to be prepared, such as labeling a danger mark at dangerous locations, or safety equipment needs to be installed, which will ensure that operators cannot put their hands at dangerous locations.
(4) Raise awareness to always be very attentive	So that people do not get used to danger, constant reminders are needed, such as pointing at certain target objects with a finger, or during the regular morning meeting on the production floor before the operation starts for the day.
(5) Improve skills and standardize work	There is a limit to people's attentiveness or reminders. → To avoid incidents caused by the lack of experience, skills and techniques, take advantage of experienced people (knowing safe and efficient ways to do the work). → Prepare a work standard in which improvements are made in terms of the 3 mu's (*muda*, *muri*, *mura* *) and raise skills.
(6) Avoid fatigue	Take care of physical conditions, such as how to avoid working while feeling fatigue after holidays.
(7) Think of team work with great care	If some people think it is fine if they are attentive to safety issues, explain that it is necessary to have each other's cooperation for safety in the work place.
(8) Make habit of following the rules or what has been decided	Accidents often happen when the rules or what has been decided are not observed. → Thorough training is needed, so as to avoid recurring accidents, and also to observe what has been agreed to as rules.

Bibliography

1) Taiichi Ohno. Toyota seisan houshiki. Diamond-Sha (Diamond Inc.), 1978.
2) Nihon Noritsu Kyokai (Japan Management Association). Toyota-no gemba-kanri. Nihon Noritsu Kyokai (Japan Management Association), 1978.
3) Katsuyoshi Ishihara. Gemba-no IE tekisuto (Book 1 & 2 or Jo & Ge). Nikka Giren Shuppansha (Union of Japanese Scientists and Engineers, or JUSE Press, Ltd.), 1978.
4) Shigeo Shingo. Toyota seisan houshiki no IE-teki shikou. Nikkan Kogyo Shimbun, Ltd., 1980.
5) Nikkan Kogyo Shimbun, Ltd. toyota-no tsuyosa-no genten – ohno taiichi-no kaizen-damashii. Nikkan Kogyo Shimbun, Ltd., 2005.
6) Ritsushi Tsukuda. shizai koubai gyoumu kanzen manual. Urbanproduce Corporation, 1997.
7) Ritsushi Tsukuda. hinshitsu kanri-ga wakaru hon. Nihon Noritsu Kyokai Management Center (JMA Management Center Inc.), 1998.
8) Ritsushi Tsukuda. furyou zero taisaku-no susume-kata. Nihon Noritsu Kyokai Management Center (JMA Management Center Inc.), 2002.

<About the Author>

Ritsushi Tsukuda, born in 1943 in Kumamoto Prefecture, Japan.
- Graduated from Tokyo University of Science, Faculty of Science.
- After working for a chemical manufacturer, Mr. Tsukuda served as Councilor with the Japan Management Association.
- He is owner of Tsukuda Sogo Keiei Jimusho, providing consultation in cost reduction, productivity improvement, defect reduction and quality management at companies (in such industries as material, chemistry, semiconductor, electronic / mechanical component parts, machine processing / assembling, consumer electronics).
- Registered SME (Small and Medium Enterprise) Business Diagnostic Engineer, or Chuushou Kigyou Shindanshi.
- ISO9000 Chief Examiner (JRCA)

<Main Publication>
- jissen tahinshu shoryou seisan no hinshitsu kanri. Nikkan Kogyo Shimbun, Ltd.
- shizai koubai gyoumu kanzen manual. Urbanproduce Corporation.
- gemba kansatsu-ho ni yoru furyou taiji. Nihon Noritsu Kyokai Management Center (JMA Management Center Inc.).
- hinshitsu kanri-ga wakaru hon. Nihon Noritsu Kyokai Management Center (JMA Management Center Inc.).
- furyou zero taisaku-no susume-kata. Nihon Noritsu Kyokai Management Center (JMA Management Center Inc.).

ALSO FROM – 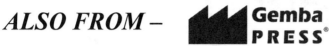

Taiichi Ohno's Workplace Management

Taiichi Ohno

Translation by Jon Miller

"This book brings to us Taiichi Ohno's philosophy of workplace management – the thinking behind the Toyota Production System. I personally get a thrill down my spine to read these thoughts in Ohno's own words. My favorite part is his discussion of the misconceptions hidden within common sense and how management needs a revolution of awareness."

- Dr. Jeffrey Liker, Director
Japan Technology Management Program, University of Michigan and Author, The Toyota Way

"Many lean students would want nothing more than to spend a day with Taiichi Ohno walking through their plant. This book is the closest thing we have left to that experience."

- Jamie Flinchbaugh, Co-author
The Hitchhiker's Guide to Lean: Lessons from the Road

"This book and its translation provide the reader a wonderful opportunity to learn directly from the master architect of the Toyota Production System. One is able to hear, in his own words, the principles that have evolved into the most successful management method ever developed. Today, these lessons are being applied in many industries including health care in addition to their long term application in manufacturing. This book enables the reader to get inside Taiichi Ohno's thinking as he makes concepts such as Kanban, The Supermarket System and Just in Time come alive in ways that can be easily understood. This book will help me, as a senior executive in health care, better implement our management method, the Virginia Mason Production System."

- Gary S. Kaplan, MD, Chairman and CEO
Virginia Mason Medical Center

AVAILABLE NOW
ww.gemba.com/press

COMING SOON –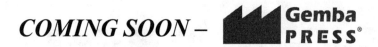

The Illustrated Toyota Production System Book2
– The Practice of Continuous Improvement

Ritsushi Tsukuda

Translation by Mark T. Nagai

The second book of this series builds on the theoretical foundation of the first book by providing 60 practical lessons on the Toyota Production System and the techniques of kaizen.

In Section 1, "What is Kaizen?" the author gives insight into what kaizen is and how to do it through practical answers such as "how can we learn to see waste?" and "what to do when you cannot find topics for kaizen?" and also insights on reducing defects and unplanned stoppages.

In section 2, "Industrial Engineering (IE) Methods are the Foundation of Kaizen at Toyota" the author introduces the most common IE methods such as process analysis, motion analysis, work sampling and line balancing.

In section 3, "Achieving Manpower Savings by Analyzing Each Process" the focus is on enabling multi-process handling by reducing wasted motion, transportation and problem solving.

In section 4, "Kaizen within Flow Lines Using Time Analysis" the emphasis is on creating a smooth flow by reducing waiting time, improving motion economy and balancing cycle time and takt time.

This book teaches immediately useful ideas and tools for continuous improvement that can be used again and again to reduce waste and improve productivity, safety and quality.

AVAILABLE SPRING 2009

www.gemba.com/press

Go See the Toyota Production System in Action
On a Gemba Tour to Japan

I hope you gained new insights from this book. It has been a great learning experience for me as well. Part of my research for preparing to translate this book was to visit many Lean companies in Japan to make sure I understood things correctly. Those trips have been invaluable in helping me translate the nuances correctly and I have found that you never stop learning!

If you are like me and would like to see what you read first hand, join me on Gemba's Japan Kaikaku Experience. It is a week filled with site tours and behind the scenes explanations, and discussions with your peers about what it means and how to apply it.

Mark. T Nagai

- Each site is carefully selected for specific areas of excellence in Lean practices
- Plant tours, Q&A sessions, and reviews with senior management reinforce the knowledge and insights gained.
- Daily review sessions link what you have seen, heard, and experienced, to Lean principles.

Scheduling Your Trip

Public JKE trips are scheduled and open for enrollment three times per year. Come by yourself or better yet bring your boss with you.

Private Trips

For groups of over 15 participants, we can customize an agenda that suits your organization's focus and timeframe.

To register for a trip or to learn more: info@gemba.com

Gemba™
An Operational Excellence Consultancy
Teaching the Toyota Production System

Since 1998 Gemba has implemented lean systems at more than 200 manufacturing, distribution and service organizations worldwide. Gemba staff are based in the North America, South America, Asia and Europe. We partner with clients for single site, multi-site and global lean deployments.

At Gemba our focus is on teaching and implementing the Toyota Production System, a constantly evolving system that encompasses people development and process innovation.

We Develop the Strategy

- Benchmarking tours
- Diagnostics & assessments
- Policy Deployment training

We Educate and Train People

- Lean systems and culture overview
- Lean leadership
- Organization design to sustain a lean culture
- Practical problem solving
- Lean facilitator certification
- *Six sigma* certification
- Various *Toyota Production System* methods and tools

We Accelerate Improvement

- *Kaizen* events
- Focused improvement projects (safety, quality, delivery, cost, capacity, ventory)
- 100-week *lean transformations*

We Implement Lean Systems

- *Safety* & ergonomics in process design and people development
- Practical *industrial engineering skills* for continuous improvement
- *Value stream mapping*
- Visual management systems
- Standard work development and Documentation
- Suggestion system design and deployment
- *Energy audits*
- Built in quality systems
- Lot size and lead-time reduction activities through *SMED* and heijunka
- Internal material logistics systems
- Team leader & group leader development
- Design processes for flow
- Improve equipment reliability through *TPM*
- Design and deploy *lean supply base* and external logistics (inbound)
- Design and deploy lean distribution and logistics strategy (outbound)
- Deploy lean to new product introduction (NPI) process
- Lean practices in knowledge work

www.gemba.com

Email: info@gemba.com
Tel:+1-425-356-3150